INSTRUCTOR'S MANUAL TO ACCOMPANY

BRADY / HUMISTON / THIRD EDITION

GENERAL CHEMISTRY

PRINCIPLES AND STRUCTURE

175 YEARS OF
1807 JW 1982
PUBLISHING

JOHN WILEY & SONS NEW YORK CHICHESTER BRISBANE TORONTO SINGAPORE

CONTENTS

ISBN 0 471 86367 X

Printed in the United States of America

10 9 8 7 6 5 4 3 2

Examinations

For a two-semester course we feel it is desirable to give at least three (and preferably four) full period exams per semester, plus a final exam at the end of each semester which covers the entire term's work. This permits adequate depth of coverage of the topics dealt with on each exam. You may also find it worthwhile to give a number of shorter quizzes throughout the term as well. (These would be given in recitation sections, if your course has them.)

Organization of the Instructor's Manual

This manual is divided into three principal sections. In the first, learning objectives for each chapter are presented along with the answers to all the end-of-chapter review questions. (Solutions to the review problems are available in a separate supplement - see below.) The second section provides a list of suggested lecture demonstrations. In the third section you will find reprinted all the stereo illustrations that appeared in the second edition of the text.

Supplements

Supplements available to accompany the text, in addition to this Instructor's Manual, include the following:

Student Study Guide, James E. Brady. The study guide is keyed section by section to the text. Each section includes objectives, review, self-test, and a list of new terms. Answers to the self-tests are given at the ends of the chapters. The study guide also includes a complete glossary.

Card File of Test Items, David Becker. A comprehensive set of multiple choice questions.

Solutions Manual, Theodore W. Sottery. Contains complete worked-out solutions to all the numerical problems that appear at the ends of the chapters.

Laboratory Manual, Jo A. Beran and James E. Brady. This includes 46 experiments plus special introductory sections on laboratory safety and techniques, liberally illustrated with photographs. Experiments of both a qualitative and quantitative nature are included. For each experiment there is a prelab assignment and a report sheet.

Instructor's Laboratory Manual, Jo A. Beran. A complete teachers manual for the laboratory manual, listing special equipment needed, amounts of chemicals, suggested unknowns, special precautions, and answers to the prelab assignments and questions on the report sheets.

Problem Exercises for General Chemistry, 2nd Ed., G. Gilbert Long and Forrest C. Hentz. This book contains over 1300 problems and questions designed to teach students to solve problems and answer questions in a format similar to that found on examinations.

Computer Aided Instruction for General Chemistry, William A. Butler. Tutorial help
for understanding chemical principles that is provided by twenty self-contained micro-
computer programs. Most programs have a "menu" of 5 to 6 parts, and each part is
usually divided into 10 or 12 entries consisting of problems, questions, etc. The
programs will initially be available in diskette format only, for the Pet, Apple II, and
TRS 80 Microcomputers.

CHAPTER 1

INTRODUCTION

Rationale

In this chapter we have provided a rather thorough introduction to basic concepts. Even though most general chemistry students have had a chemistry course in high school, the lack of uniformity of depth of coverage and of retained knowledge makes it impossible to assume any particular level of training to serve as a point of departure. Therefore, we have assumed the student mind to be a blank slate. The coverage that you give to the topics in this chapter has to be governed by your appraisal of your students' backgrounds. If students can come away from this chapter with an understanding of its contents, they should have a firm foundation upon which to build the remainder of the course.

Objectives

After completing the chapter students should be able to:

Differentiate between law and theory; qualitative and quantitative observations; precision and accuracy; mass and weight; extensive and intensive properties; physical and chemical properties; homogeneous and heterogeneous; exothermic and endothermic.

Apply the concepts of significant figures and exponential notation in carrying out mathematical computations using the factor-label method.

Perform conversions among units in the metric/SI system.

Differentiate between substances that are either elements, compounds or mixtures.

Explain the laws of definite proportions, conservation of mass and multiple proportions by applying Dalton's Atomic Theory.

Write the chemical symbols of the common elements.

Perform computations involving density and specific gravity.

Give the number of atoms specified in a chemical formula.

Determine whether or not a chemical equation is balanced.

Associate changes in potential energy with changes in attractive and repulsive forces.

Define kinetic energy.

Perform conversions among the Fahrenheit, Celsius and Kelvin temperature scales.

Perform simple calculations relating to specific heat.

Answers to Questions

1.1   A law is a generalized statement of fact that has been obtained by experiment. Theories attempt to explain how nature behaves in the operation of laws.

1.2   Significant figures indicate the reliability of measurements.

1.3   Precision refers to how closely two measurements of the same quantity come to each other while accuracy refers to how close an experimental observation lies to the actual value.

1.4   The units gram and milliliter are more appropriate in the laboratory because of the size (small) of the quantities used.

1.5   An extensive property is one that depends on the size of the sample used. An intensive property is one that is independent of the size of the sample used.

Extensive Properties	Intensive Properties
Force (weight), length, number of atoms	Freezing point, color, specific heat

1.6   Density is the ratio of an object's mass to its volume. Specific gravity is the ratio of the object's density to the density of water. Units for density are g/ml; specific gravity has no units.

1.7   Homogeneous   Heterogeneous
       sea water   smog
       air   smoke
       black coffee   club soda (with bubbles)
       penny   ham sandwich

1.8   There are 4 phases: copper pan, iron nails, glass marbles, water.

1.9   The mass of an object does not vary from place to place; it is the same regard-
       less of where it is measured.

1.10   Elements are the simplest forms of matter that can exist under ordinary chemi-
       cal conditions. Compounds consist of two or more elements. Mixtures consist
       of two or more compounds that do not react chemically.

1.11   Atomic mass.

1.12   Atoms are the fundamental particles of all matter that cannot be further sub-
       divided by ordinary chemical means. A molecule is a group of atoms bound
       tightly enough together that they behave as one.

1.13   (a) Fe (b) Na (c) K (d) P (e) Br (f) Ca (g) N (h) Ne (i) Mn (j) Mg

1.14   (a) silver   (d) chlorine   (g) chromium   (j) mercury
       (b) copper   (e) aluminum   (h) tungsten
       (c) sulfur   (f) gold   (i) nickel

1.15   (a) 1 K, 2 S   (d) 3 N, 12 H, 1 P, 4 O
       (b) 2 Na, 1 C, 3 O   (e) 3 Na, 1 Ag, 4 S, 6 O
       (c) 4 K, 1 Fe, 6 C, 6 N

1.16 $CaSO_4 \cdot 2H_2O$

1.17   (a) During a chemical reaction, mass is neither created nor destroyed.

       (b) In any pure chemical substance, the same elements are always combined in
            the same proportions, by mass.

       (c) When two elements combine to form more than one compound, the masses
            of one of the elements that are combined with the same mass of the other
            in the various compounds are in ratios of small whole numbers.

1.18   The amu is one-twelfth of the mass of an atom of carbon-12.

1.19 (a) and (c)

1.20 <u>Potential energy</u> is the energy due to position of particles that either attract or repel each other. <u>Kinetic energy</u> is the energy due to motion; $KE = \frac{1}{2}mv^2$

1.21 An endothermic process absorbs heat (energy) as it proceeds. An exothermic process emits heat (energy) as it proceeds.

1.22 Repulsive

1.23 <u>Heat</u> is a form of energy, whereas the measure of the intensity of heat is known as <u>temperature</u>.

1.24 cal/g °C or J/g °C. For water, sp. ht. = 1.000 cal/g °C or 4.184 J/g °C. Water has the largest specific heat of any common substance. During the winter the water loses some of its heat and prevents the surrounding land masses from becoming too cold. During the summer, the water can absorb large amounts of heat and prevent surrounding land masses from becoming too hot.

CHAPTER 2

STOICHIOMETRY: CHEMICAL ARITHMETIC

Rationale

Stoichiometry is presented here so that you may begin quantitative laboratory experiments early in the course. You will find our introduction to the mole concept quite different from that found in most other texts. This approach has worked very well with students, particularly in getting them to think in terms of mole ratios when they look at subscripts in formulas and coefficients in equations. In this edition we have included a new section in this chapter that deals with molar concentration and explains how solutions of a desired molarity can be prepared. If you wish, at this time you can also cover additional topics on solution stoichiometry that are found in Section 6.9 (Pages 193-196). Appropriate homework problems on this topic can be assigned by referring to the Index to Questions and Problems on Page 205.

Objectives

Upon completion of this chapter, students should be able to:

Apply the factor-label method in stoichiometry calculations.

Interpret subscripts in formulas and coefficients in equations in terms of mole ratios.

Convert grams of a substance to moles, and vice versa.

Calculate molecular weight (formula weight) from a chemical formula.

Calculate percentage composition.

Calculate an empirical formula from percentage composition or data from a chemical analysis.

Determine the molecular formula from an empirical formula and the molecular weight of the substance.

Balance a chemical equation by inspection, and interpret the equation on a mole basis.

Use a balanced chemical equation to perform chemical calculations.

Determine the limiting reactant given the quantities of two or more reactants.

Calculate theoretical yield and percentage yield if given the actual yield.

Calculate the molarity of a solution given the amount of solute and the volume of the solution.

Explain how a solution having a desired molarity should be prepared.

Use molarity as a conversion factor in calculating (a) the amount of solute in a known volume of a solution of a given molarity, (b) the volume of a solution of known molarity that is required to contain a specified amount of solute, and (c) the amount of solute needed to prepare a given volume of solution having a specified molarity.

Answers to Questions

2.1   All three represent a set number of objects; the mole is 6.02×10^{23} things, the dozen is 12 things and the gross means 144 things.

2.2   Formula weights are preferred in compounds where formula units are used to describe a neutral aggregate of ions, e.g. $NaCl$.

2.3   Structural formulas represent the way the atoms in a molecule are linked together. Molecular formulas specify the actual numbers of each kind of atom found in a molecule. Empirical formulas give the relative numbers of atoms of each element present in a molecule.

2.4 NH_4SO_4, Fe_2O_3, $AlCl_3$, CH, $C_3H_8O_3$, CH_2O, Hg_2SO_4

2.5   molecular formula is $C_2H_6O_2$; empirical formula is CH_3O.

2.6   A simplest formula is calculated from experimentally observed or measured data obtained by a chemical analysis of the compound. Webster's defines empirical as "Pertaining to, or founded upon, experiment or experience."

INTRODUCTION

The text, General Chemistry: Principles and Structure, was written for a full year General Chemistry course for science students. These would include students majoring in agriculture, biology, chemistry, engineering, geology, pharmacy and the allied health professions, physics, and those preparing for medical, dental, or veterinary school.

The scope and emphasis of general chemistry courses at different colleges and universities vary greatly. Textbooks naturally reflect this by trying to incorporate sufficient material so that they can be adapted to a variety of educational situations. This text covers all of the topics any ordinary science student might be expected to learn in general chemistry. This does not mean, of course, that everything in the book must necessarily be presented. After reviewing the detailed table of contents you may wish to de-emphasize certain topics, depending on the specific educational needs of your students, their prior chemical training, and their overall intellectual potential.

The sequence of topics in the text has changed somewhat from one edition to another and reflects both the authors' bias with respect to the order of presentation, as well as input from users of the previous editions. Surely there are other equally valid orders of presentation, and in general you will find that topics are treated so that their order can be easily modified to suit your own preferences. For example, in this edition we have moved the second bonding chapter forward so that it now occurs immediately following the first one, because that is where the majority of users want it. However, this chapter, and those that follow, are written so that instructors who wish to treat the more advanced bonding topics during the second semester can easily do so.

Descriptive Chemistry

An area of particular concern to teachers of chemistry in recent times has been the subject of descriptive chemistry. In this third edition we have completely overhauled our treatment of this material to provide both greater depth of coverage, as well as greater flexibility for the instructor in deciding how much descriptive chemistry he or she wishes to present.

In addition to the descriptive chemistry woven into discussions, worked examples, and end-of-chapter questions and problems, there are now three principal loca-

1

tions where descriptive chemistry is concentrated. The first of these is in Chapter 6 (Chemical Reactions in Aqueous Solutions), where ionic reactions and redox reactions are discussed. We consider this to be especially important for all students, and in our classes we cover this chapter thoroughly.

The second place where descriptive chemistry is concentrated is in Chapter 9 (The Periodic Table Revisited). Its location - following chapters on bonding and the properties of the states of matter - permits a meaningful discussion of some important trends in the physical and chemical properties of the elements. Although the location of the chapter places it toward the end of the first semester, the chapter is sufficiently portable that it could easily be presented later in the course.

In courses that place a heavy emphasis on "Principles," instructors may wish to limit their discussion of descriptive chemistry to Chapters 6 and 9. However, if a more detailed treatment of the chemistry of the elements is desired, additional topics can be chosen from Chapters 18 through 21. In these chapters the goal has been to describe the principal compounds and reactions of the most important elements. As you will see, there have been frequent references to familiar substances and applications of chemistry.

Scheduling

Since the division of the academic year into semesters, trimesters, quarters, etc., varies so much from school to school, we feel it would be best to simply provide you with our estimate of the number of 50-minute periods needed to cover the subject matter in the various chapters. A little arithmetic will reveal that the total number of hours probably exceeds the number that you have available. Therefore, as mentioned earlier, you will have to be somewhat selective in choosing which chapters you will cover completely.

Chapter	Periods	Chapter	Periods
1	3	13	3
2	4	14	3
3	4	15	5
4	4	16	2
5	3	17	4
6	5	18	2
7	4	19	2
8	4	20	3
9	3	21	4
10	4	22	3
11	4	23	3
12	4	24	2

2.7   The law of conservation of mass.

2.8   Coefficients are: (a) 1,2,1,1 (b) 6,2,2,3 (c) 8,3,4,9 (d) 1,1,2 (e) 2,1,2,1

2.9   Coefficients are: (a) 2,3,1,6 (b) 2,1,1,2,2 (c) 3,2,1,6 (d) 1,3,1,3,3
       (e) 1,1,1,1,2

2.10   Coefficients are:
       (a) 2, 13, 8, 10 (1, 13/2, 4, 5)   (d) 4, 11, 2, 8 (2, 11/2, 1, 4)
       (b) 2, 17, 14, 6 (1, 17/2, 7, 3)   (e) 4, 5, 4, 6 (2, 5/2, 2, 3)
       (c) 1, 6, 4

2.11   That reactant that is completely consumed before the remaining reactants are
       used up. First calculate the number of moles of each reactant present. Then
       compare their ratios with that for the balanced chemical equation. From this
       deduce which reactant will be depleted first.

2.12   The theoretical yield is the maximum amount of product that could be produced
       from a given quantity of reactant regardless of any other products. The per-
       cent yield is a comparison of the theoretical yield to the yield that is actually
       obtained; it is the efficiency of the reaction. The actual yield is the amount
       of product that you actually obtain in a given experiment when the reaction is
       carried out.

2.13 $$\text{molar concentration} = \frac{\text{number of moles of solute}}{\text{total volume of the solution in liters}}$$

2.14   Place 180 g of $C_6H_{12}O_6$ in a 1.00-liter volumetric flask. Dissolve the sugar in
       some water; then dilute to a total volume of 1.00 liter.

2.15   0.20 M Na_3PO_4 means
       $$\frac{0.20 \text{ mol } Na_3PO_4}{1000 \text{ ml solution}} \quad \text{and} \quad \frac{1000 \text{ ml solution}}{0.20 \text{ mol } Na_3PO_4}$$

CHAPTER 3

ATOMIC STRUCTURE AND THE PERIODIC TABLE

Rationale

This chapter introduces students to the detailed structure of matter for the first time. In this chapter one of our goals has been to show students how theory must relate to experimentally observable phenomena. For example, Sections 3.1 through 3.7 examine, historically, the development of our current picture of the atom. The structure of the periodic table and the periodic law are discussed because the theory relating to electronic structure must explain the structure of the periodic table. The Bohr theory is examined because it was successful at accounting for the line spectrum of hydrogen. We also see here how theories must be abandoned when they fail. Another reason for discussing the Bohr theory is to show how a theory relating to atomic structure is checked against "reality," i.e. by deriving an equation from the theory that matches with one derived empirically (Page 80).

The remainder of the chapter deals with the modern view of atomic structure and electron configuration, as well as the variation of some properties with variations in electronic structure. A more qualitative treatment of this chapter might include a condensation of Sections 3.1 to 3.7, with the elimination of numerical calculations. You may also wish to limit the discussion of the Bohr theory to simply an examination of the Bohr model of the atom and its reason for failure.

Objectives

At the completion of this chapter students should be able to:

Relate the contributions of the following scientists toward our picture of the atom: Faraday, Thomson, Millikan, Becquerel, Rutherford, Moseley, Chadwick, Bohr, Planck, Einstein, DeBroglie and Schrödinger.

Define isotopes and compute the average atomic weight of an element given the relative abundances and actual masses of its isotopes.

Perform calculations relating wavelength and frequency for electromagnetic radiation.

Relate the contributions of Mendeleev and Meyer to the periodic classification of the elements.

For the periodic table, identify: groups and periods; representative, transition and inner transition elements; rare earth elements; alkali metals; alkaline earth metals; halogens; noble gases; metals; nonmetals; metalloids.

Explain the difference between a continuous spectrum and a line spectrum.

Explain how an atomic spectrum is obtained experimentally.

Give the relationship between energy and frequency for a photon.

Describe the Bohr model of the atom.

Explain how a diffraction pattern is formed.

Identify the four quantum numbers and specify their allowed values.

State the Pauli exclusion principle and Hund's rule.

Use the periodic table to predict the electron configurations of the elements.

Use the location of a representative element in the periodic table to write the electron configuration of its outer shell.

Describe the shapes of s and p atomic orbitals.

Describe and explain the periodic trends in the following properties: atomic size, ionic size, ionization energy, and electron affinity.

Answers to Questions

3.1   Cathode rays travel in straight lines, cast shadows, turn pinwheels, heat metal foil, and can be bent by an electric or magnetic field.

3.2   A coulomb is the amount of electric charge moving past a given point in a wire when an electric current of one ampere flows for one second.

3.3   See Figure 3.1 on Page 62.

3.4   Electrons (cathode rays) passing through the gas knock electrons off neutral molecules, which leaves positively charged particles behind.

3.5   Canal rays are rays of positive particles that pass through a perforated cathode.

3.6   Hydrogen is the lightest of all elements and has the largest e/m ratio for any positive ion.

3.7   Alpha, α, He^{2+} particles; beta, β, composed of electrons; gamma, γ, rays, high energy light waves.

3.8   The path of the particles is guided by a magnetic field which deflects them into a circular path. The degree of curvature of these paths is determined by the charge-to-mass ratio of the ions.

3.9   Because some of the alpha particles were strongly deflected by the thin foil.

3.10   See Table 3.1, Page 68.

3.11

	protons	neutrons	electrons
$^{132}_{55}Cs$	55	77	55
$^{115}_{48}Cd^{2+}$	48	67	46
$^{194}_{81}Tl$	81	113	81
$^{105}_{47}Ag^{1+}$	47	58	46
$^{78}_{34}Se^{2-}$	34	44	36

3.12

	protons	neutrons	electrons
$^{131}_{56}Ba$	56	75	56
$^{109}_{48}Cd^{2+}$	48	61	46
$^{36}_{17}Cl^{-}$	17	19	18

$^{63}_{28}$Ni   28   35   28

$^{170}_{69}$Tm   69   101   69

3.13   (a) $^{55}_{26}$Fe   (b) $^{86}_{37}$Rb   (c) $^{204}_{81}$Tl   (d) $^{170}_{71}$Lu   (e) $^{169}_{70}$Yb

3.14   (a) 29   (b) 49   (c) 123   (d) 89   (e) 99

3.15 $^{12}_{6}$C

3.16   Because the observed atomic weights are obtained as an average of the masses contributed by each isotope of an atom.

3.17   The mass number is simply the total count of protons plus neutrons and is not quite equal to the atomic mass of an atom.

3.18   When the elements are arranged in order of increasing atomic number, there occurs a periodic repetition of similar chemical and physical properties.

3.19   To leave room for yet undiscovered elements.

3.20   Elements arranged in order of increasing atomic weights. Elements with similar properties placed in columns.

3.21   Atomic weight = 71, Melting point = 409°C, Boiling point = 2233°C,
Formula of chloride, $GaCl_3$
Formula of oxide, Ga_2O_3
Melting point of chloride, 388°C
Atomic weight and formulas agree very well. Boiling point is high, as expected. Other properties do not agree very well.

3.22

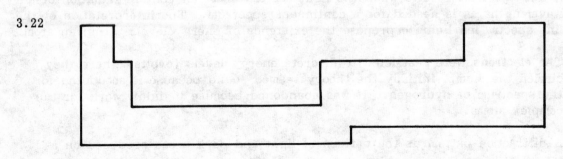

3.23   Mg, Se and Br

3.24   Ru, W and Ag

3.25   Elements 58 through 71 and 90 through 103.

3.26   F

3.27   K

3.28   Ba

3.29   Ta, Nd and Cs

3.30   Nonmetals are poor conductors of heat and electricity and lack luster.

3.31   The chemical properties of the elements are determined by the way in which the electrons are arranged. The structure of the table has to be explained by the theory.

3.32   See Figure 3.10 on Page 73. Wavelength, λ, is the distance between consecutive peaks in a wave. Frequency, ν, is the number of peaks passing a given point per second. They are related to each other by the equation $\lambda \cdot \nu = c$ where c is the speed of light.

3.33   SI unit of frequency is the hertz, $1 \text{ Hz} = 1 \text{ s}^{-1}$. Units for wavelengths are chosen so that the numbers are simple to comprehend. Thus, 320 nm is easier to comprehend than 3.20×10^{-7} m. The visible region of the spectrum runs from about 400 nm to 800 nm.

3.34   Infrared light has longer wavelengths than visible light; ultraviolet light has shorter wavelengths than visible light.

3.35   A line spectrum results when the light emitted does not contain radiation of all wavelengths as is needed for a continuous spectrum. The interpretation of the line spectra led Bohr to propose the existence of energy levels within an atom.

3.36   The electrons were restricted to discrete energy levels (orbits) where they circled the atom. Initially the theory seemed sound because it accounted for the spectrum of hydrogen. It was abandoned because it didn't work for more complex atoms.

3.37   A diffraction pattern is the pattern of light and dark areas produced on a screen when two beams of diffracted light interact with each other by constructive and destructive interference.

3.38 Because the wavelengths are too small to be detected.

3.39 Electrons and other subatomic particles can be used to produce diffraction patterns.

3.40 A wave whose nodes are stationary.

3.41 $n = 1, 2, 3, 4, \ldots, \infty$
$\ell = 0, 1, 2, \ldots, (n - 1)$
$m = 0, \pm 1, \pm 2, \ldots, \pm \ell$

3.42 The ground state is the state of lowest energy.

3.43 s = 2; p = 6; d = 10; f = 14; g = 18; h = 22
First shell with an h subshell ($\ell = 5$) has n = 6.
For an h subshell, $m = 0, \pm 1, \pm 2, \pm 3, \pm 4, \pm 5$

3.44 Pauli exclusion principle - see Page 87.
Hund's rule - see Page 91.

3.45 For a filled L shell:

electron no.	n	ℓ	m	s
1	2	0	0	1/2
2	2	0	0	-1/2
3	2	1	1	1/2
4	2	1	1	-1/2
5	2	1	0	1/2
6	2	1	0	-1/2
7	2	1	-1	1/2
8	2	1	-1	-1/2

3.46 18

3.47 (a) P

↑↓	↑↓	↑↓ ↑↓ ↑↓	↑↓	↑ ↑ ↑
1s	2s	2p	3s	3p

(b) Ca

↑↓	↑↓	↑↓ ↑↓ ↑↓	↑↓	↑↓ ↑↓ ↑↓	↑
1s	2s	2p	3s	3p	4s

3.48   Sc   [Ar]   ↑↓   ↑ __ __ __ __   (P)

Ti   [Ar]   ↑↓   ↑ ↑ __ __ __   (P)

V   [Ar]   ↑↓   ↑ ↑ ↑ __ __   (P)

Cr   [Ar]   ↑   ↑ ↑ ↑ ↑ ↑   (P)

Mn   [Ar]   ↑↓   ↑ ↑ ↑ ↑ ↑   (P)

Fe   [Ar]   ↑↓   ↑↓ ↑ ↑ ↑ ↑   (P)

Co   [Ar]   ↑↓   ↑↓ ↑↓ ↑ ↑ ↑   (P)

Ni   [Ar]   ↑↓   ↑↓ ↑↓ ↑↓ ↑ ↑   (P)

Cu   [Ar]   ↑   ↑↓ ↑↓ ↑↓ ↑↓ ↑↓   (P)

Zn   [Ar]   ↑↓   ↑↓ ↑↓ ↑↓ ↑↓ ↑↓   (D)

3.49   Predicted electron configurations based on position in periodic table.

P; $1s^2 2s^2 2p^6 3s^2 3p^3$

Ni; $1s^2 2s^2 2p^6 3s^2 3p^6 4s^2 3d^8$

As; $1s^2 2s^2 2p^6 3s^2 3p^6 4s^2 3d^{10} 4p^3$

Ba; $1s^2 2s^2 2p^6 3s^2 3p^6 4s^2 3d^{10} 4p^6 5s^2 4d^{10} 5p^6 6s^2$

Rh; $1s^2 2s^2 2p^6 3s^2 3p^6 4s^2 3d^{10} 4p^6 5s^2 4d^7$

Ho; $1s^2 2s^2 2p^6 3s^2 3p^6 4s^2 3d^{10} 4p^6 5s^2 4d^{10} 5p^6 6s^2 5d^1 4f^{10}$

Sn; $1s^2 2s^2 2p^6 3s^2 3p^6 4s^2 3d^{10} 4p^6 5s^2 4d^{10} 5p^2$

3.50   Si, $3s^2 3p^2$; Se, $4s^2 4p^4$; Sr, $5s^2$; Cl, $3s^2 3p^5$; O, $2s^2 2p^4$; S, $3s^2 3p^4$; As, $4s^2 4p^3$; Ga, $4s^2 4p^1$.

3.51   Rb $1s^2 2s^2 2p^6 3s^2 3p^6 4s^2 3d^{10} 4p^6 5s^1$

Sn $1s^2 2s^2 2p^6 3s^2 3p^6 4s^2 3d^{10} 4p^6 5s^2 4d^{10} 5p^2$

Br $1s^2 2s^2 2p^6 3s^2 3p^6 4s^2 3d^{10} 4p^5$

Cr $1s^2 2s^2 2p^6 3s^2 3p^6 4s^1 3d^5$

Cu $1s^2 2s^2 2p^6 3s^2 3p^6 4s^1 3d^{10}$

3.52 K $4s^1$; Al $3s^2 3p^1$; F $2s^2 2p^5$; S $3s^2 3p^4$; Tl $6s^2 6p^1$; Bi $6s^2 6p^3$

3.53 In a Bohr orbit, the electron is confined to a fixed path about the nucleus. In an orbital, the electron is free to move anywhere in the atom. The concept of the probability of locating the electron within a minute volume element at various places applies to orbitals and is a consequence of the uncertainty principle.

3.54 See Figure 3.29.

3.55 The size increases from 1s to 2s etc. and nodes occur. Their overall shape however is spherical.

3.56 The s orbital is spherical while the p orbital is dumbbell-shaped.

3.57 A node is a position where the amplitude of a wave is zero. For an electron wave, it is a place where the probability of finding the electron is zero.

3.58 Three electrons, which have the same charge, will tend to stay away from each other as far as possible. This they can do by occupying separate p orbitals.

3.59 An atom or ion has no fixed outer limits. Atomic and ionic sizes are usually given in angstroms, nanometers, or picometers.

3.60 Ge

3.61 (a) Se (b) C (c) Fe^{2+} (d) O^- (e) S^{2-}

3.62 $r_{N^{3-}} = 1.71$ Å, $r_{O^{2-}} = 1.40$ Å, $r_{F^-} = 1.36$ Å. These ions are isoelectronic (i.e., they have identical electron configurations). The effective nuclear charge is increasing N < O < F.

3.63 The lanthanide contraction is the gradual decrease in the sizes of the lanthanide elements that occurs upon the filling of the inner 4f subshell in the lanthanides. As a result, Hf is nearly the same size as Zr, but with a much larger nuclear charge pulling on the outermost electrons. Outer electrons are held more tightly.

3.64 When there are fewer electrons, the interelectron repulsions are less and the outer shell can contract in size under the influence of the nuclear charge.

3.65 Ionization energy is the amount of energy needed to remove an electron from an atom or ion.

Electron affinity is the amount of energy released or absorbed when an electron is added to an atom or an ion.

3.66　As we move from left to right across a period the increased effective nuclear charge causes the shell to shrink in size and also makes it more difficult to remove an electron.

3.67　(a) Be　(b) Be　(c) N　(d) N　(e) Ne　(f) S^+　(g) Na^+

3.68　This graph is supposed to illustrate the stability of the noble gas electron configuration.

3.69　(a) Cl　(b) S　(c) P (predicted from general trends; actually, this is an exception and EA for As is more exothermic.)　(d) S

3.70　The second electron that is added must be forced into an already negative ion. This requires work and therefore is an endothermic process.

3.71　The Group VIIA elements can attain a stable configuration by acquiring one electron. The octet is a much more stable configuration than just seven electrons and when the Group VIIA elements do achieve their octet a relatively large amount of energy is released.

3.72
Faraday	-	electrical nature of matter
Thomson	-	charge-to-mass ratio of the electron
Rutherford	-	nuclear atom
Millikan	-	charge on the electron
Moseley	-	nuclear charge (atomic number) from X-ray data
Schrödinger	-	developed wave equation and founded wave mechanics
de Broglie	-	matter has wave properties
Planck	-	energy of a photon = $h\nu$

CHAPTER 4

CHEMICAL BONDING: GENERAL CONCEPTS

Rationale

Chemical bonding is divided between two chapters. In this chapter we discuss the "traditional" concepts of ionic and covalent bonding, using Lewis structures to keep track of valence electrons. Chapter 5 takes a closer look at molecular structure and the more modern theories of bonding based on wave mechanics.

As stated, the intent of Chapter 4 is to provide an elementary introduction to chemical bonding. We have also introduced oxidation-reduction and oxidation numbers here because they follow quite naturally from the concepts of ionic bond formation and polarity of molecules. The introduction of oxidation numbers also serves as a foundation for a discussion of nomenclature of inorganic compounds, which is included within the text in this edition, rather than in an appendix.

Objectives

Upon completion of this chapter, students should be able to:

Write the Lewis symbols for the representative elements.

Distinguish between ionic and covalent bonding.

Construct Lewis structures to illustrate ionic bonding.

Write the formulas of ionic compounds.

Explain the factors that influence the formation of ionic compounds using a Born-Haber cycle; define lattice energy.

Construct Lewis structures for molecules and polyatomic ions.

Define bond order and understand its relationship to bond length, bond energy, and bond vibrational frequency.

Predict when resonance is important in describing the bonding in simple molecules and construct appropriate Lewis structures.

Illustrate the concept of coordinate covalent bonding.

Apply the concept of electronegativity to explain dipole moment.

Describe trends in electronegativity within the periodic table.

Distinguish between oxidation and reduction; oxidizing agents and reducing agents.

Assign oxidation numbers or oxidation states to the atoms in molecules and ions.

Name inorganic compounds according to the rules in Section 4.10.

Answers to Questions

4.1 $:\overset{\bullet}{\underset{\bullet\bullet}{Se}}\bullet$, $:\overset{\bullet}{\underset{\bullet\bullet}{Br}}\bullet$, $\bullet\,Al\,\bullet$, $K\bullet$, $\bullet\,Ba\bullet$, $\bullet\,\overset{\bullet}{Ge}\,\bullet$ and $:\overset{\bullet}{\underset{\bullet}{P}}\bullet$

4.2   Lewis symbols help us to keep tabs on the outer shell electrons which are most important in bond formation.

4.3   Cation is a positively charged ion; anion is a negatively charged ion.

4.4 $Ba^{2+}\ [:\overset{\bullet\bullet}{\underset{\bullet\bullet}{O}}:]^{2-}$; $2Na^{+}, [:\overset{\bullet\bullet}{\underset{\bullet\bullet}{O}}:]^{2-}$; $K^{+}[:\overset{\bullet\bullet}{\underset{\bullet\bullet}{F}}:]^{-}$; $Mg^{2+}, 2[:\overset{\bullet\bullet}{\underset{\bullet\bullet}{F}}:]^{-}$

4.5   KF(s) is stable because the compound has a very large lattice energy, the ionization energy of K is low and the electron affinity of F is high. The lattice energy is primarily responsible for the stability of the compound.

4.6   Looking at outer-shell configurations

K	$4s^1 \longrightarrow 4s^0$	O	$2s^2 2p^4 \longrightarrow 2s^2 2p^6$
Mg	$3s^2 \longrightarrow 3s^0$	N	$2s^2 2p^3 \longrightarrow 2s^2 2p^6$
Na	$3s^1 \longrightarrow 3s^0$	S	$3s^2 3p^4 \longrightarrow 3s^2 3p^6$
Ba	$6s^2 \longrightarrow 6s^0$	Br	$4s^2 4p^5 \longrightarrow 4s^2 4p^6$

4.7 $Ba^{2+}[Xe]$, $Se^{2-}[Kr]$, $Al^{3+}[Ne]$, $Li^{+}[He]$, $Br^{-}[Kr]$, $Fe^{2+}[Ar]3d^{6}$, $Cu^{+}[Ar]3d^{10}$, $Ni^{2+}[Ar]3d^{8}$

4.8   The second shell can contain a maximum of 8 electrons.

4.9 $ns^{2}np^{6}nd^{10}$, Zn^{2+}, Cd^{2+}, Hg^{2+}

4.10   (a) Na_2CO_3 (b) $Ca(ClO_3)_2$ (c) SrS (d) $CrCl_3$ (e) $Ti(ClO_4)_4$

4.11   (a) $Fe_2(HPO_4)_3$ (b) K_3N (c) $Ni(NO_3)_2$ (d) $Cu(C_2H_3O_2)_2$ (e) $BaSO_3$

4.12   Because for many of them, their outer shells have two electrons.

4.13   An element in periods 4, 5, or 6 that is also in Group IIIA to VIIA.

4.14

$$Li(s) \;+\; 1/2\, Br_2(\ell) \longrightarrow LiBr(s)$$

endothermic: 1,2,3

exothermic: 4,5

4.15   Lattice Energy. Exothermic

4.16 $\cdot \ddot{N} \cdot \;+\; 3H\cdot \longrightarrow$ H:N̈:H with H below

$\cdot \ddot{O} \cdot \;+\; 2H\cdot \longrightarrow$ H:Ö:H

$\cdot \ddot{F}: \;+\; H\cdot \longrightarrow$ H:F̈:

4.17   Cl—P—Cl (with Cl below), H—Si—H (with H above and below), Cl—B—Cl (with Cl above),

H—S—H (bent), H—C—C—C—H (with H above and below each C), :C≡O:

24

4.18
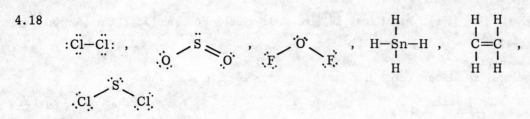

4.19
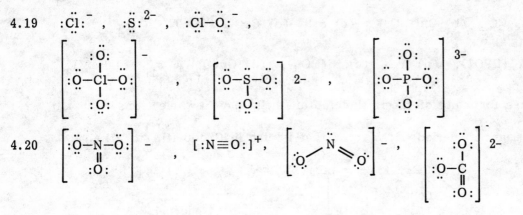

4.20
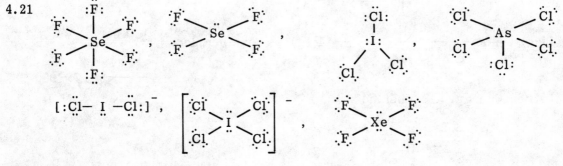

4.21

4.22 ClF_3, SF_4, IF_7, BCl_3

4.23 1.54 Å, 1.46 Å, 1.40 Å and 1.37 Å. Bond energy also increases in this direction.

4.24

Molecule	CO Bond Order	CO Bond Length	CO Bond Energy
CO	3		
CO_2	2		
CH_3COO^-	1.5	increases	decreases
CO_3^{2-}	1.3		
CH_3CH_2OH	1		

4.25 Vibrational frequencies, which increase with increasing bond order.

4.26 A resonance hybrid is the true structure of a compound and is a composite of the contributing structures. We use resonance because it is impossible to draw a single electron-dot formula that obeys the octet rule and is consistent with experimental facts at the same time.

4.27

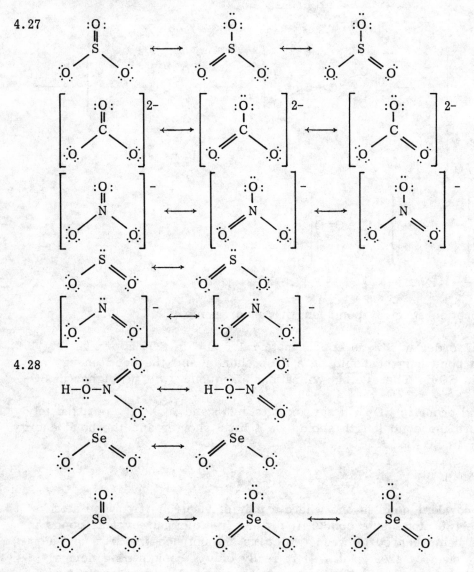

4.28

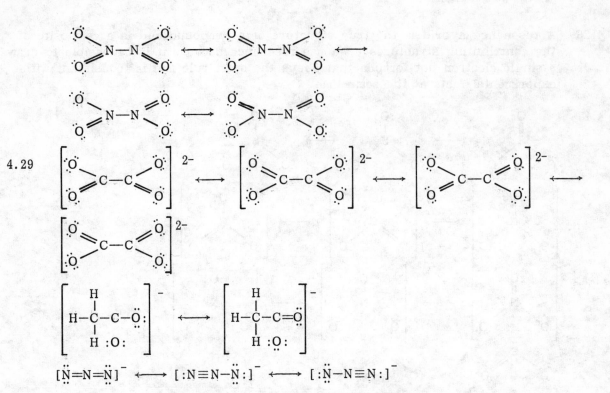

4.30 Infrared absorption spectra, bond length measurements

4.31 The S—O bond order in SO_2 is approximately 1.5 and in SO_3 approximately 1.3. Therefore the bond length should be a little shorter and the bond energy a little higher in SO_2. The average vibrational frequency will be higher in SO_2.

4.32 The N—O bond order in NO_2^- is approximately 1.5 and in NO_3^- approximately 1.3. Therefore the bond length should be a little shorter and the bond energy a little higher in NO_2^-.

4.33 Bond order = 1.3

4.34 A coordinate covalent bond is one where a pair of electrons from one atom is shared by the two atoms in a bond. It differs from normal covalent bonds in that electrons being shared between two atoms are both coming from one atom instead of one electron from each. It is really only a bookkeeping device; once formed the coordinate covalent bond is the same as any other covalent bond.

4.35
$$
\begin{array}{c}
:\ddot{C}l: \\
| \\
:\ddot{C}l-Al \\
| \\
:\ddot{C}l:
\end{array}
\;+\;
[\;{}^{x}_{x}\ddot{C}l{}^{x}_{x}\;]^{-}
\;\longrightarrow\;
\left[
\begin{array}{c}
:\ddot{C}l: \\
| \\
:\ddot{C}l-Al\leftarrow Cl\,{}^{x}_{x}{}^{x}_{x} \\
| \\
:\ddot{C}l:
\end{array}
\right]^{-}
$$

4.36   Dative bond

4.37   Electronegativity is the attraction an atom has for electrons in a chemical bond. Electron affinity, which is an energy term referring to an isolated atom, is the energy released or absorbed when an electron is added to a neutral gaseous atom.

4.38   A polar molecule is one that has its positive and negative charges separated by a distance. The resulting molecule is said to be a dipole. The dipole moment is the product of the charge on either end of the dipole times the distance between the charges.

4.39   Fluorine, upper right-hand corner, is the element highest in electronegativity. The values generally decrease down a group and right to left across a period. Elements with low ionization energies generally have low electronegativities and those with high ionization energies also have high electronegativities.

4.40 MgO, Al_2O_3, CsF

4.41 NH_3, BCl_3, BeI_2, NaH

4.42 F_2, H_2Se, H_2S, OF_2, SO_2, ClF_3, and SF_2

4.43   Oxidation is the process of losing electrons. Reduction is the process of gaining electrons. Oxidation state is the charge an atom would have if both electrons in the bond were assigned to the more electronegative element. An oxidizing agent is that material being reduced (causing oxidation). A reducing agent is that material being oxidized (causing reduction).

4.44   K Cl O_2,   Ba Mn O_4, Fe_3 O_4, O_2 F_2,   I F_5
        1+ 3+ 2-   2+ 6+ 2-   8/3+ 2-   1+ 1-   5+ 1-

        H O Cl,   Ca S O_4, Cr_2 (S $O_4)_3$,
        1+ 2- 1+   2+ 6+ 2-   3+   6+ 2-

        O_3, Hg_2 Cl_2
        0   1+   1-

4.45 Ox. No. of C

C_2H_5OH $CH_3C{\overset{O}{\underset{}{\diagdown}}}{-}H$ $CH_3C{\overset{O}{\underset{}{\diagdown}}}{-}OH$ CO_2

  (-2)   (-1)   (0)   (+4)

4.46 H_2 S O_4,   C   Br,   O F_2, H_2 O_2,   Cr Cl_3, Mn_2 O_7

      1+   6+   2-   4+   1-   2+   1-   1+   1-   3+   1-   7+   2-

     K   Mn O_4, H_2 C_2 O_4,   K   Cl O_3,   Li   N O_3

      1+   7+   2-   1+   3+   2-   1+   5+   2-   1+   5+   2-

4.47   (a) oxidation   (b) reduction   (c) oxidation   (d) oxidation   (e) reduction

4.48

NaBr	sodium bromide
CaO	calcium oxide
$FeCl_3$	ferric chloride; iron(III) chloride
$CuCO_3$	cupric carbonate; copper(II) carbonate
CBr_4	carbon tetrabromide
P_4O_6	tetraphosphorushexaoxide; phosphorus(III) oxide
$AsCl_5$	arsenic pentachloride; arsenic(V) chloride
$Mn(HCO_3)_2$	manganous hydrogen carbonate; manganous bicarbonate
	manganese(II) hydrogen carbonate; manganese(II) bicarbonate

4.49 $Al(NO_3)_3$, $FeSO_4$, $NH_4H_2PO_4$, IF_5, PCl_3, N_2O_4, $KMnO_4$, $Mg(OH)_2$, H_2Se, NaH

4.50

Chromium(III) oxide	Aluminum phosphate
Magnesium dihydrogen phosphate	Magnesium nitride
Copper(II) nitrate	Lead(II) oxalate
Calcium sulfate	Ammonium carbonate
Barium hydroxide	Potassium dichromate

4.51

TiO_2	$Ni(HCO_3)_2$
$SiCl_4$	$NaHSO_4$
CaSe	$(NH_4)_2Cr_2O_7$
KNO_3	$Ca(C_2H_3O_2)_2$
$Al_2(SO_4)_3$	$Sr(OH)_2$

CHAPTER 5

COVALENT BONDING AND MOLECULAR STRUCTURE

Rationale

In previous editions, this chapter occurred in the second half of the book. Its location there made sense because students would be exposed to these topics after they had had time to digest the simpler concepts, and because reference to the material covered here could easily be avoided until discussions of descriptive chemistry later in the course. It also makes sense, of course, to present this chapter immediately following the basic bonding concepts, because the two chapters back-to-back provides a continuous, logical development of concepts. Most users of the text prefer the present order, but the chapter has been written so that those who wish to continue to present these concepts during the second semester can easily do so.

After describing the basic geometries common to nearly all molecular structures, the VSEPR theory is presented. In applying the theory, it is important, of course, that students have learned to draw Lewis structures. The VSEPR theory is followed by discussions of valence bond and molecular orbital theories.

In comparing the various theories, an important point to make is that although they seem different, they are all describing the same phenomenon. None of the theories is perfect, but each is useful in its own way. The VSEPR is marvelous in its simplicity and its effectiveness in predicting molecular structure. Valence bond theory permits the retention of the simple picture of the electron-pair bond, being formed by overlap of appropriate orbitals. Molecular orbital theory is useful once molecular structure is known because it allows us to discuss the energy levels in molecules and because it avoids resonance.

Objectives

At the conclusion of this chapter, students should be able to:

Describe the basic molecular shapes: linear, planar triangular, tetrahedral, trigonal bipyramidal, and octahedral.

State the basic postulates on which the VSEPR, valence bond, and molecular orbital theories are based.

Apply the VSEPR theory to predict the shapes of simple molecules and polyatomic ions.

Use the valence bond theory to explain the bonding and shapes of various molecules and polyatomic ions.

Describe the orientations of the various hybrid orbitals.

Use the VSEPR to help provide a valence bond description of a molecule or polyatomic ion.

Describe the shapes of bonding and antibonding MO's formed by orbitals of the first and second shells in homonuclear diatomic molecules.

Use the basic concepts of the molecular orbital theory to explain the electron distribution and the bonding in H_2, Li_2, Be_2, B_2, C_2, N_2, O_2 and F_2; to explain why He_2 and Ne_2 do not exist; and to explain why O_2 is a paramagnetic molecule.

Answers to Questions

5.1   See the figures on Pages 140-142.

5.2   180°

5.3   planar triangular, 120°; tetrahedral, 109.5°; octahedral, 90°.

5.4   Within triangular plane, 120°; between triangular plane and apex, 90°.

5.5   Covalent bonds between any bonded species consist of a sharing of electron pairs. The geometric arrangements of atoms, or groups of atoms about some central atom is determined solely by the mutual repulsion between the electron pairs (both bonding and lone pairs) present in the valence shell of the central atom.

5.6   (a) AX_3E; pyramidal   (f) AX_2E_3; linear
     (b) AX_4; tetrahedral   (g) AX_5E; square pyramidal
     (c) AX_3; planar triangular   (h) AX_4; tetrahedral
     (d) AX_2E; V-shaped   (i) AX_6; oxtahedral
     (e) AX_4E; distorted tetrahedral   (j) AX_5; trigonal bipyramidal

(k) AX_2E; V-shaped

5.7   (1)   (2)
   (a)   planar triangular   planar triangular
   (b)   tetrahedral   pyramidal
   (c)   octahedral   octahedral
   (d)   tetrahedral   pyramidal
   (e)   tetrahedral   pyramidal
   (f)   tetrahedral   tetrahedral
   (g)   planar triangular   planar triangular
   (h)   tetrahedral   tetrahedral
   (i)   tetrahedral   nonlinear (bent)
   (j)   octahedral   square planar

5.8   (1)   (2)
   (a)   tetrahedral   bent
   (b)   tetrahedral   bent
   (c)   octahedral   octahedral
   (d)   tetrahedral   tetrahedral
   (e)   octahedral   square planar
   (f)   tetrahedral   tetrahedral
   (g)   planar triangular   planar triangular
   (h)   trigonal bipyramidal   T-shaped
   (i)   planar triangular   planar triangular
   (j)   tetrahedral   pyramidal
   (k)   tetrahedral   tetrahedral

5.9   (a)   planar triangular to tetrahedral
   (b)   trigonal bipyramidal to octahedral
   (c)   T-shaped to square planar
   (d)   nonlinear (V-shaped) to distorted tetrahedral
   (e)   linear to "double" bent

   (f)   planar triangular to linear

5.10   When two atoms come together to form a covalent bond, an atomic orbital of one atom overlaps with an atomic orbital of the other, and a pair of electrons is shared between the two atoms in the region of overlap.

5.11   The p orbital with one electron on one Cl atom overlaps with the p orbital with one electron on the other Cl atom.

5.12 Sn has the valence shell configuration, $5s^2 5p^2$. If the Cl atoms' orbitals overlap with the p orbitals of Sn, a nonlinear (bent) molecule should result with a bond angle of about 90° (actual Cl—Sn—Cl angle in $SnCl_2$ is 95°).

5.13 Pure p orbitals from the As $(4s^2 4p^3)$.

5.14 A bond angle of 90° is predicted if S uses two unhybridized p orbitals. The slightly greater bond angle can be explained on the basis of repulsions between the H atoms.

5.15 P $\underset{3s}{\uparrow\downarrow}$ $\underset{3p}{\uparrow \quad \uparrow \quad \uparrow}$

F $\underset{2s}{xx}$ $\underset{2p}{xx \quad xx \quad x}$

P (in PF_3) $\underset{3s}{\uparrow\downarrow}$ $\underset{3p}{\uparrow x \quad \uparrow x \quad \uparrow x}$

$\cdot \ddot{P} \cdot$, $\cdot \ddot{F} :$

The shape would be pyramidal.

5.16 Hybrid orbitals are produced when two or more atomic orbitals are mixed, producing a new set of orbitals.

5.17 We must employ hybrid orbitals in order to account for the H—C—H bond angle of 109° in CH_4 and the equivalence of the four C—H bonds.

5.18 (a) sp^2 (b) sp^3 (c) dsp^3 (d) d^2sp^3 (e) sp (f) d^2sp^3 (g) sp^3 (h) dsp^3 (i) sp^3

5.19 (a) planar triangular   (f) octahedral
     (b) tetrahedral   (g) pyramidal
     (c) trigonal bipyramidal   (h) distorted tetrahedral
     (d) octahedral   (i) tetrahedral
     (e) linear

5.20 (a) sp^3 (b) sp^3 (c) sp^2 (d) sp^2 (e) $sp^3 d$ (f) $sp^3 d$ (g) $sp^3 d^2$ (h) sp^3 (i) $sp^3 d^2$ (j) $sp^3 d$ (k) sp^2

5.21 (a) sp^2 (b) sp^3 (c) $sp^3 d^2$ (d) sp^3 (e) sp^3 (f) sp^3 (g) sp^2 (h) sp^3 (i) sp^3 (j) $sp^3 d^2$

5.22   Sb   ↓↑   ↑   ↑   ↑   — — — — —
        5s   5p   5d

x = Cl electrons

$SbCl_5$ ↑x   ↑x ↑x ↑x   ↑x   — — — — —

$\underbrace{\qquad\qquad\qquad\qquad}_{dsp^3}$ unhybridized 5d orbitals

5.23   Si has a relatively low-energy d subshell in its valence shell, whereas C does not.

5.24   Boron in BCl_3 is probably sp^2 hybridized. Nitrogen in NH_3 is sp^3 hybridized. In order for boron to accept two electrons from the nitrogen in Cl_3BNH_3, it must become sp^3 hybridized.

5.25   A coordinate covalent bond "exists" in NH_4^+, $AlCl_6^{3-}$, $SbCl_6^-$ and ClO_4^-. After formation, the coordinate covalent bond is (of course) identical to the "normal" covalent bond.

5.26   (a) 109.5° (b) 120° (c) 180° (d) 90°

5.27   Sn is sp^3 hybridized. The Sn—Cl bond would be formed by sp^3-p overlap.

5.28   The σ-bond results from a head-on overlap of atomic orbitals that concentrates electron density along the imaginary line joining the bound nuclei. A π-bond is produced by the sideways overlap of atomic p orbitals providing electron density above and below the line connecting the bound nuclei. A double bond consists of one σ-bond and one π-bond. A triple bond is usually made by overlap of orbitals to give one σ and two π-bonds.

5.29

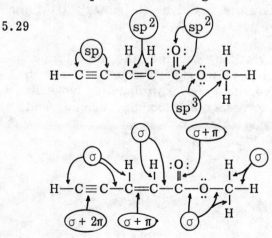

5.30 (a)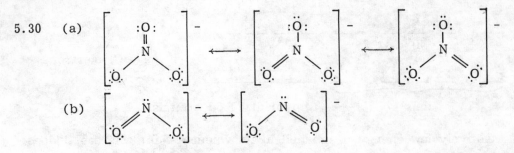

(b)

5.31 Molecules are created by the formation of a set of molecular orbitals that are spread over the nuclei of the molecule. These molecular orbitals are formed by atomic orbital overlap. Orbital overlap is important in both VB and MO theories. However, valence bond permits us to retain our picture of individual atoms coming together to form a covalent bond. Resonance, which is needed in VB theory to explain bonding in some molecules, is not needed in MO theory.

5.32 Valence bond theory would show that the N_2 triple bond consists of one p-p σ and two p-p π-bonds. The molecular orbital picture of N_2, shown in Figure 5.22 shows that the triple bond consists of doubly occupied σ_{2p_z}, π_{2p_x} and π_{2p_y} molecular orbitals. Both views of bonding yield the same result.

5.33 The molecular orbital diagram of N_2 can be seen in Figure 5.22. The species N_2^+ would have one less π_{2p} electron and N_2^- would have an additional electron located in the π_{2p}^* molecular orbital. Both N_2^+ and N_2^- are less stable than N_2 and both would have a longer bond length than N_2.

5.34 The molecular orbital diagram of O_2 can be seen in Figure 5.22. The species O_2^+ would have one less π_{2p}^* electron and the species O_2^- would have one additional π_{2p}^* electron. The stability therefore would be $O_2^+ > O_2 > O_2^-$; the bond length would increase in the same order.

5.35 A bonding molecular orbital is obtained by adding together the wave functions of the two overlapping atomic orbitals; antibonding molecular orbitals are obtained by subtracting these same two wave functions. Antibonding molecular orbitals are higher in energy than their corresponding bonding orbitals and, as a result, destabilize a molecule when they are filled.

5.36

	Li$_2$	Be$_2$	B$_2$	C$_2$
$\sigma^*_{2p_z}$	—		—	—
$\pi^*_{2p_x}\ \pi^*_{2p_y}$	— —	— —	— —	— —
$\pi_{2p_x}\ \pi_{2p_y}$	— —	— —	— —	↑ ↑
σ_{2p_z}	—	—	↑↓	↑↓
σ^*_{2s}	—	↑↓	↑↓	↑↓
σ_{2s}	↑↓	↑↓	↑↓	↑↓

Actually, for these species the σ_{2p} and π_{2p} levels are reversed in order. However, for simplicity in teaching the principles of MO theory, this has been ignored in the text. The MO energy level diagram on Page 166 has been used to answer this question.

5.37   (a)   Because a bonding electron is removed in each case, Li_2^+, B_2^+, and C_2^+ would be less stable than the neutral X_2 species. Be_2^+ would be more stable than Be_2 because an antibonding electron is removed.

     (b) Li_2^- would be less stable, Be_2^-, B_2^-, and C_2^- would be more stable than the neutral X_2 species. The extra electron is antibonding in Li_2^- but is bonding in Be_2^-, B_2^-, and C_2^-.

5.38   It allows for the formation of molecular orbitals that extend over more than two nuclei. This delocalization is the equivalent of VB resonance.

5.39 H_2^+ Bond order = 1/2; $(\sigma_{1s})^1$

      He_2^+ Bond order = 1/2; $(\sigma_{1s})^2(\sigma^*_{1s})^1$

5.40   <u>Valence Bond Description</u>: In all three species the central atom is sp^2 hybridized. In SO_2 two resonance structures arise from alternately creating a double bond between the unhybridized p orbital on S and p orbitals on the oxygens. In NO_3^- three resonance structures arise because there are three O atoms which can π-bond to the unhybridized p orbital on N. In HCO_2^- the C—H bond is a single bond, and two resonance structures result by forming two possible double bonds as in the case of SO_2.
<u>Molecular Orbital Description</u>: Again, the central atom is sp^2 hybridized in all three cases. Delocalized π clouds are formed over three atoms in both SO_2 and HCO_2^- (i.e., over O—S—O and over O—C—O). In NO_3^- a delocalized π cloud extends over all 4 atoms as in the case of SO_3 (Figure 5.23, Page 167).

5.41   Figure 5.3 (b) and (c). X—A—X angles will be compressed to less than 109.5° in perfect tetrahedron (e.g., H_2O and NH_3).

      Figure 5.4 (b) distorted tetrahedral (arrows indicate direction of distortion).

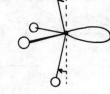

      (c) T-shape

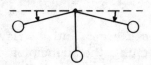

      Figure 5.5 (b) square pyramid

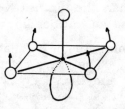

5.42   In NH_3, the N—H bonds are polar with the negative ends of the bond dipoles at the nitrogen. The lone pair, if it were in a nonbonded sp^3 orbital, would also produce a contribution to the dipole moment of the molecule, with its negative end pointing <u>away</u> from the nitrogen.

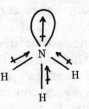

effects of the bond dipoles
and lone pair dipole are additive

All dipoles are additive and produce a large net molecular dipole.

In NF_3, the fluorines are more electronegative than nitrogen, producing bond dipoles with their positive ends at nitrogen. These, therefore, tend to offset the contribution of the lone pair in the sp^3 orbital on nitrogen, thereby giving a very small net dipole moment for the NF_3 molecule.

bond dipoles cancel the effects
of the lone pair dipole

CHAPTER 6

CHEMICAL REACTIONS IN AQUEOUS SOLUTION

Rationale

This chapter, at an early point in the book, treats many topics that are useful in laboratory experiments. The chapter contains a large amount of very basic, important descriptive chemistry that students must know for future courses in chemistry and for other disciplines. Students should learn the material thoroughly.

Objectives

Upon completion of this chapter students should be able to:

Distinguish between the terms: solvent and solute; concentrated and dilute; saturated, unsaturated, and supersaturated; electrolyte and nonelectrolyte.

Explain what is meant by a dynamic equilibrium.

Write molecular, ionic, and net ionic equations for ionic reactions, and balance them.

Apply the solubility rules on Page 179.

Describe the factors that lead to a net ionic reaction.

Recognize and name the common inorganic acids and bases.

Classify acids and bases as monoprotic or polyprotic, strong or weak and write the equilibrium equation for ionization of the weak acids and bases in Table 6.3.

Write chemical equations for neutralization reactions.

Devise metathesis reactions for the preparation of some inorganic salts.

Balance oxidation-reduction equations by both the oxidation-number-change method and the ion-electron method.

Calculate the molarity of solutions.

Carry out calculations involving solution stoichiometry.

Compute equivalent weights.

Use the concepts of equivalent weight and normality to carry out calculations involving the stoichiometry of solutions.

Use the concepts in the chapter to solve problems involving gravimetric and volumetric analyses.

Perform calculations dealing with dilution.

Answers to Questions

6.1   Water is one of the most abundant chemicals in nature and serves as a good solvent for many substances, both ionic and molecular. Because of the homogeneous nature of solutions, dissolved substances are intimately mixed and chemical reactions occur rapidly.

6.2   Solvent: substance present in the greatest proportion in a solution
       Solutes: all other substances present
       Concentrated: relatively large amounts of solute present in the solvent
       Dilute: only a small amount of solute in the solvent
       Saturated: at a given temperature, the maximum amount of solute that the solvent can hold in contact with undissolved solute
       Supersaturated: at a given temperature, the solvent contains more solute than it can usually hold in the presence of excess solute
       Unsaturated: the solvent contains less than the maximum amount at a given temperature

6.3   Solubility: the amount of solute required to produce a saturated solution at a particular temperature

6.4   We can distinguish between strong and weak electrolytes with the aid of the apparatus shown in Figure 6.3. Ions are present in solutions of electrolytes but not in solutions of non-electrolytes.

6.5 $KCl \longrightarrow K^+ + Cl^-$

$(NH_4)_2SO_4 \longrightarrow 2NH_4^+ + SO_4^{2-}$

$Na_3PO_4 \longrightarrow 3Na^+ + PO_4^-$

$NaOH \longrightarrow Na^+ + OH^-$

$HCl \longrightarrow H^+ + Cl^-$

6.6 H^+ is the short form of H_3O^+. We often leave out the H_2O which is merely a carrier for the H^+ ion.

6.7   A dynamic equilibrium is one in which opposing processes occur at equal rates, so there is no <u>net</u> change in the system, e.g. ions react to form molecules while molecules react to form ions. $CdSO_4 \rightleftharpoons Cd^{2+} + SO_4^-$

6.8 $H_2O \rightleftharpoons H^+ + OH^-$ (or, $2H_2O \rightleftharpoons H_3O^+ + OH^-$)

6.9   By position of equilibrium, we mean the relative proportions of reactants and products in a chemical system when the system is at equilibrium.

6.10   It lies mainly to the left, in favor to undissociated H_2O. For HCl the position of equilibrium lies virtually 100% in favor of the ionic products.

6.11   Precipitate

6.12   Metathesis is the kind of reaction in which the cations and anions have changed partners. This is also known as double replacement.

6.13 $HCN \rightleftharpoons H^+ + CN^-$

$H_2S \rightleftharpoons H^+ + HS^-$

$HS^- \rightleftharpoons H^+ + S^{2-}$

$NH_3 + H_3O^+ \rightleftharpoons NH_4^+ + H_2O$

$HgCl_2 \rightleftharpoons HgCl^+ + Cl^-$

$HgCl_2 \rightleftharpoons Hg^{2+} + 2Cl^-$

6.14   (a) $Cu^{2+} + 2Cl^- + Pb^{2+} + 2NO_3^- \longrightarrow Cu^{2+} + 2NO_3^- + PbCl_2(s)$

         Net ionic $Pb^{2+}(aq) + 2Cl^-(aq) \longrightarrow PbCl_2(s)$

      (b) $Fe^{2+} + SO_4^{2-} + 2Na^+ + 2OH^- \longrightarrow Fe(OH)_2(s) + 2Na^+ + SO_4^{2-}$

         Net ionic $Fe^{2+}(aq) + 2OH^-(aq) \longrightarrow Fe(OH)_2(s)$

(c) $Zn^{2+} + SO_4^{2-} + Ba^{2+} + 2Cl^- \longrightarrow Zn^{2+} + 2Cl^- + BaSO_4(s)$

Net ionic $Ba^{2+}(aq) + SO_4^{2-}(aq) \longrightarrow BaSO_4(s)$

(d) $2Ag^+ + 2NO_3^- + 2K^+ + SO_4^{2-} \longrightarrow Ag_2SO_4(s) + 2K^+ + 2NO_3^-$

Net ionic $2Ag^+(aq) + SO_4^{2-}(aq) \longrightarrow Ag_2SO_4(s)$

(e) $2NH_4^+ + CO_3^{2-} + Ca^{2+} + 2Cl^- \longrightarrow 2NH_4^+ + 2Cl^- + CaCO_3(s)$

Net ionic $Ca^{2+}(aq) + CO_3^{2-}(aq) \longrightarrow CaCO_3(s)$

6.15 (a) $Al(OH)_3(s) + 3H^+ + 3Cl^- \longrightarrow Al^{3+} + 3Cl^- + 3H_2O$

Net ionic $Al(OH)_3(s) + 3H^+(aq) \longrightarrow Al^{3+}(aq) + 3H_2O$

(b) $CuCO_3(s) + 2H^+ + SO_4^{2-} \longrightarrow Cu^{2+} + SO_4^{2-} + H_2O + CO_2(g)$

Net ionic $CuCO_3(s) + 2H^+ \longrightarrow Cu^{2+}(aq) + H_2O + CO_2(g)$

(c) $Cr_2(CO_3)_3(s) + 6H^+ + 6NO_3^- \longrightarrow 2Cr^{2+} + 6NO_3^- + 3H_2O + 3CO_2(g)$

Net ionic $Cr_2(CO_3)_3(s) + 6H^+ \longrightarrow 2Cr^{3+}(aq) + 3H_2O + 3CO_2(g)$

(d) $3Cu(s) + 8H^+ + 8NO_3^- \longrightarrow 3Cu^{2+} + 6NO_3^- + 2NO(g) + 4H_2O$

Net ionic $3Cu(s) + 8H^+ + 2NO_3^-(aq) \longrightarrow 3Cu^{2+}(aq) + 2NO(g) + 4H_2O$

(e) $MnO_2(s) + 4H^+ + 4Cl^-(aq) \longrightarrow Mn^{2+} + 2Cl^- + Cl_2(g) + 2H_2O$

Net ionic $MnO_2(s) + 4H^+(aq) + 2Cl^-(aq) \longrightarrow Mn^{2+} + Cl_2(g) + 2H_2O$

6.16

Soluble	Insoluble
KCl	$CaCO_3$
$(NH_4)_2SO_4$	$PbSO_4$
$AgNO_3$	$Mn(OH)_2$
$Zn(ClO_4)_2$	$Fe\ PO_4$
$Ba(C_2H_3O_2)_2$	NiO

6.17 (a) $Ag^+ + Br^- \longrightarrow AgBr(s)$

(b) $CoCO_3(s) + 2H^+ \longrightarrow Co^{2+} + CO_2(g) + H_2O$

(c) $C_2H_3O_2^- + H^+ \longrightarrow HC_2H_3O_2$

(d) $Pb^{2+} + SO_4^{2-} \longrightarrow PbSO_4(s)$

(e) $H_2S + Cu^{2+} \longrightarrow 2H^+ + CuS(s)$

(f) $NH_4^+ + OH^- \longrightarrow NH_3 + H_2O$

6.18 (a) $Na_2SO_4 + BaCl_2 \longrightarrow BaSO_4(s) + 2NaCl$

$2Na^+ + SO_4^{2-} + Ba^{2+} + 2Cl^- \longrightarrow BaSO_4(s) + 2Na^+ + 2Cl^-$

$Ba^{2+} + SO_4^{2-} \longrightarrow BaSO_4(s)$

(b) $Ca(NO_3)_2 + (NH_4)_2CO_3 \longrightarrow CaCO_3 + 2NH_4NO_3$

$Ca^{2+} + 2NO_3^- + 2NH_4^+ + CO_3^{2-} \longrightarrow CaCO_3 + 2NH_4^+ + 2NO_3^-$

$Ca^{2+} + CO_3^{2-} \longrightarrow CaCO_3$

(c) $NaC_2H_3O_2 + HNO_3 \longrightarrow NaNO_3 + HC_2H_3O_2$

$Na^+ + C_2H_3O_2^- + H^+ + NO_3^- \longrightarrow Na^+ + NO_3^- + HC_2H_3O_2$

$H^+ + C_2H_3O_2^- \longrightarrow HC_2H_3O_2$

(d) $2NaOH + CuCl_2 \longrightarrow 2NaCl + Cu(OH)_2(s)$

$2Na^+ + 2OH^- + Cu^{2+} + 2Cl^- \longrightarrow 2Na^+ + 2Cl^- + Cu(OH)_2(s)$

$Cu^{2+} + 2OH^- \longrightarrow Cu(OH)_2(s)$

(e) $(NH_4)_2CO_3 + 2HNO_3 \longrightarrow 2NH_4NO_3 + H_2O + CO_2$

$2NH_4^+ + CO_3^{2-} + 2H^+ + 2NO_3^- \longrightarrow 2NH_4^+ + 2NO_3^- + H_2O + CO_2$

$2H^+ + CO_3^{2-} \longrightarrow H_2O + CO_2$

6.19 (a) No reaction

(b) $2NH_4Br + MnSO_4 \longrightarrow (NH_4)_2SO_4 + MnBr_2$

$2NH_4^+ + 2Br^- + Mn^{2+} + SO_4^{2-} \longrightarrow 2NH_4^+ + SO_4^{2-} + Mn^{2+} + 2Br^-$

No net ionic reaction

(c) $K_2S + 2HC_2H_3O_2 \longrightarrow 2KC_2H_3O_2 + H_2S$

$2K^+ + S^{2-} + 2HC_2H_3O_2 \longrightarrow 2K^+ + 2C_2H_3O_2^- + H_2S$

$2HC_2H_3O_2 + S^{2-} \longrightarrow H_2S + C_2H_3O_2^-$

(d) $MgSO_4 + 2LiOH \longrightarrow Li_2SO_4 + Mg(OH)_2(s)$

$Mg^{2+} + SO_4^{2-} + 2Li^+ + 2OH^- \longrightarrow 2Li^+ + SO_4^{2-} + Mg(OH)_2(s)$

$Mg^{2+} + 2OH^- \longrightarrow Mg(OH)_2(s)$

(e) $AgC_2H_3O_2 + KCl \longrightarrow AgCl(s) + KC_2H_3O_2$

$Ag^+ + C_2H_3O_2^- + K^+ + Cl^- \longrightarrow AgCl(s) + K^+ + C_2H_3O_2^-$

$Ag^+ + Cl^- \longrightarrow AgCl(s)$

6.20   (a) $AgBr + KI \longrightarrow AgI + KBr$

$AgBr + K^+ + I^- \longrightarrow AgI + K^+ + Br^-$

$AgBr + I^- \longrightarrow AgI + Br^-$

      (b) $SO_2 + H_2O + BaCl_2 \longrightarrow BaSO_3 + 2HCl$

$SO_2 + H_2O + Ba^{2+} + 2Cl^- \longrightarrow BaSO_3(s) + 2H^+ + 2Cl^-$

$SO_2 + H_2O + Ba^{2+} \longrightarrow BaSO_3(s) + 2H^+$

      (c) $Na_2C_2O_4 + 2HCl \longrightarrow 2NaCl + H_2C_2O_4$

$2Na^+ + C_2O_4^- + 2H^+ + 2Cl^- \longrightarrow 2Na^+ + 2Cl^- + H_2C_2O_4$

$2H^+ + C_2O_4^{2-} \longrightarrow H_2C_2O_4$

      (d) $\begin{cases} K_2SO_3 + 2HCl \longrightarrow 2KCl + H_2SO_3 \\ H_2SO_3 \longrightarrow H_2O + SO_2 \end{cases}$

$\begin{cases} 2K^+ + SO_3^{2-} + 2H^+ + 2Cl^- \longrightarrow 2K^+ + 2Cl^- + 2H^+ + SO_3^{2-} \\ 2H^+ + SO_3^{2-} \longrightarrow H_2O + SO_2(g) \end{cases}$

$SO_3^{2-} + 2H^+ \longrightarrow H_2O + SO_2(g)$

      (e)   No reaction

6.21   An acid is any substance that increases the concentration of hydronium ion in a solution. A base is any substance that increases the hydroxide ion in a solution.

6.22   Monoprotic: $HX + H_2O \rightleftharpoons H_3O^+ + X^-$

      Diprotic: $H_2Y + H_2O \rightleftharpoons H_3O^+ + HY^-$

                      $HY^- + H_2O \rightleftharpoons H_3O^+ + Y^{2-}$

      Triprotic: $H_3Z + H_2O \rightleftharpoons H_3O^+ + H_2Z^-$

                      $H_2Z^- + H_2O \rightleftharpoons H_3O^+ + HZ^{2-}$

                      $HZ^{2-} + H_2O \rightleftharpoons H_3O^+ + Z^{3-}$

6.23   (a) $HCl + KOH \longrightarrow KCl + H_2O$

      (b) $H_2SO_4 + 2NaOH \longrightarrow Na_2SO_4 + 2H_2O$

      (c) $2H_3AsO_4 + 3Ba(OH)_2 \longrightarrow Ba_3(AsO_4)_2 + 6H_2O$

6.24 $(H^+) \quad + \quad :\overset{..}{\underset{..}{O}}:H^- \longrightarrow H:\overset{..}{\underset{..}{O}}:H$

6.25 Acid salts are the products of partial neutralization of a polyprotic acid. Examples include $NaHCO_3$, $NaHSO_4$, and Na_2HPO_4.

6.26 A base turns litmus blue. Therefore, the element would be classified as a metal, because metal oxides yield bases on reaction with water.

6.27 One possible set of reactions is:

(a) $Ba(C_2H_3O_2)_2 + (NH_4)_2SO_4 \longrightarrow BaSO_4(s) + 2NH_4C_2H_3O_2$

(b) $3FeCl_2 + 2H_3PO_4 \longrightarrow Fe_3(PO_4)_2(s) + 6HCl$

(c) $Cu(NO_3)_2 + Na_2CO_3 \longrightarrow CuCO_3(s) + 2NaNO_3$

(d) $2NaOH + MgSO_4 \longrightarrow Na_2SO_4 + Mg(OH)_2(s)$

(e) $(NH_4)_2SO_4 + Pb(NO_3)_2 \longrightarrow 2NH_4NO_3 + PbSO_4(s)$

6.28 One possible set of reactions is:

(a) See (d) above.

(b) $MnCl_2 + Ba(OH)_2 \longrightarrow Mn(OH)_2 + BaCl_2$

(c) $FeCl_2 + Pb(C_2H_3O_2)_2 \longrightarrow Fe(C_2H_3O_2)_2 + PbCl_2(s)$

(d) $NiCl_2 + Pb(ClO_4)_2 \longrightarrow PbCl_2(s) + Ni(ClO_4)_2$

(e) $(NH_4)SO_3 + BaCl_2 \longrightarrow 2NH_4Cl + BaSO_3(s)$

6.29 One possible set of reactions is:

(a) $CaCO_3 + 2HCl \longrightarrow CaCl_2 + CO_2 + H_2O$

(b) $MnCO_3 + 2HClO_4 \longrightarrow Mn(ClO_4)_2 + H_2O + CO_2$

(c) $BaSO_3 + H_2SO_4 \longrightarrow BaSO_4 + H_2O + SO_2$

(d) $NaOH + NH_4NO_3 \longrightarrow NaNO_3 + H_2O + NH_3$

(e) $(NH_4)_2CO_3 + 2HC_2H_3O_2 \longrightarrow 2NH_4C_2H_3O_2 + H_2O + CO_2$

6.30 (a) $Ca(OH)_2 + 2HNO_3 \longrightarrow Ca(NO_3)_2 + 2H_2O$

(b) $2NaOH + H_2C_2O_4 \longrightarrow Na_2C_2O_4 + 2H_2O$

(c) $NaOH + H_2SO_4 \longrightarrow H_2O + NaHSO_4$

(d) $Al_2O_3 + 6HClO_4 \longrightarrow 2Al(ClO_4)_3 + 3H_2O$

(e) $NiO + 2HBr \longrightarrow NiBr_2 + H_2O$

6.31   (a) $Cu(NO_3)_2 + 2NaOH \longrightarrow Cu(OH)_2(s) + 2NaNO_3$

            $Cu(OH)_2(s) + HCl \longrightarrow CuCl_2 + H_2O$

        (b) $BaBr_2 + Na_2SO_3 \longrightarrow BaSO_3(s) + 2NaBr$

            $BaSO_3 + 2HCl \longrightarrow BaCl_2 + H_2O + SO_2$

        (c) $Na_2SO_4 + Ba(ClO_4)_2 \longrightarrow 2NaClO_4 + BaSO_4$

        (d) $MgCl_2 + Pb(C_2H_3O_2)_2 \longrightarrow Mg(C_2H_3O_2)_2 + PbCl_2$

        (e) $Na_2SO_3 + Ba(OH)_2 \longrightarrow 2NaOH + BaSO_3(s)$

            $2NaOH + H_2CO_3(CO_2 + H_2O) \longrightarrow Na_2CO_3 + 2H_2O$

6.32   (a)   10,4,4,3,1
        (b)   10,2,1,6
        (c)   3,2,8,2,2,11,3
        (d)   3,2,3,2,4
        (e)   4,1,2,4

6.33   (a) $3Cu + 8HNO_3 \longrightarrow 3Cu(NO_3)_2 + 2NO + 4H_2O$

        (b) $MnO_2 + 4HBr \longrightarrow Br_2 + MnBr_2 + 2H_2O$

        (c) $3(CH_3)_2CHOH + 2CrO_3 + 3H_2SO_4 \longrightarrow 3(CH_3)_2CO + Cr_2(SO_4)_3 + 6H_2O$

        (d) $3PbO_2 + 2Sb + 2NaOH \longrightarrow 3PbO + 2NaSbO_2 + H_2O$

        (e) $3NO_2 + H_2O \longrightarrow 2HNO_3 + NO$

6.34   (a) HNO_3 (b) KNO_3 (c) $Na_2Cr_2O_7$ (d) HNO_3 (e) O_2

6.35   (a) $8H^+ + 2NO_3^- + 3Cu \longrightarrow 2NO + 3Cu^{2+} + 4H_2O$

        (b) $10H^+ + NO_3^- + 4Zn \longrightarrow NH_4^+ + 4Zn^{2+} + 3H_2O$

        (c) $2Cr + 6H^+ \longrightarrow 2Cr^{3+} + 3H_2$

        (d) $8H^+ + Cr_2O_7^{2-} + 3H_3AsO_3 \longrightarrow 2Cr^{3+} + 4H_2O + 3H_3AsO_4$

        (e) $10H^+ + SO_4^{2-} + 8I^- \longrightarrow 4I_2 + H_2S + 4H_2O$

        (f) $4H_2O + 8Ag^+ + AsH_3 \longrightarrow H_3AsO_4 + 8Ag + 8H^+$

        (g) $H_2O + S_2O_8^{2-} + HNO_2 \longrightarrow NO_3^- + 2SO_4^{2-} + 3H^+$

        (h) $4H^+ + MnO_2 + 2Br^- \longrightarrow Mn^{2+} + Br_2 + 2H_2O$

        (i) $2S_2O_3^{2-} + I_2 \longrightarrow 2I^- + S_4O_6^{2-}$

        (j) $IO_3^- + 3HSO_3^- \longrightarrow I^- + 3SO_4^{2-} + 3H^+$

6.36 (a) $8H^+ + Cr_2O_7^{2-} + 3CH_3CH_2OH \longrightarrow 2Cr^{3+} + 3CH_3CHO + 7H_2O$

(b) $4H^+ + PbO_2 + 2Cl^- \longrightarrow Pb^{2+} + Cl_2 + 2H_2O$

(c) $14H^+ + 2Mn^{2+} + 5BiO_3^- \longrightarrow 2MnO_4^- + 5Bi^{3+} + 7H_2O$

(d) $3H_2O + ClO_3^- + 3HAsO_2 \longrightarrow 3H_3AsO_4 + Cl^-$

(e) $2H_2O + PH_3 + 2I_2 \longrightarrow H_3PO_2 + 4I^- + 4H^+$

(f) $16H^+ + 2MnO_4^- + 10S_2O_3^{2-} \longrightarrow 5S_4O_6^{2-} + 2Mn^{2+} + 8H_2O$

(g) $4H^+ + 2Mn^{2+} + 5PbO_2 \longrightarrow 2MnO_4^- + 5Pb^{2+} + 2H_2O$

(h) $2H^+ + As_2O_3 + 2NO_3^- + 2H_2O \longrightarrow 2H_3AsO_4 + N_2O_3$

(i) $8H_2O + 2P + 5Cu^{2+} \longrightarrow 5Cu + 2H_2PO_4^- + 12H^+$

(j) $6H^+ + 2MnO_4^- + 5H_2S \longrightarrow 2Mn^{2+} + 5S + 8H_2O$

6.37 (a) $H_2O + CN^- + AsO_4^{3-} \longrightarrow AsO_2^- + CNO^- + 2OH^-$

(b) $2CrO_2^- + 3HO_2^- \longrightarrow 2CrO_4^- + H_2O + OH^-$

(c) $7OH^- + 4Zn + NO_3^- + 6H_2O \longrightarrow 4Zn(OH)_4^{2-} + NH_3$

(d) $4OH^- + Cu(NH_3)_4^{2+} + S_2O_4^{2-} \longrightarrow 2SO_3^{2-} + Cu + 4NH_3 + 2H_2O$

(e) $N_2H_4 + 2Mn(OH)_3 \longrightarrow 2Mn(OH)_2 + 2NH_2OH$

(f) $4H_2O + 2MnO_4^- + 3C_2O_4^{2-} \longrightarrow 2MnO_2 + 6CO_2 + 8OH^-$

(g) $6OH^- + 7ClO_3^- + 3N_2H_4 \longrightarrow 6NO_3^- + 7Cl^- + 9H_2O$

6.38 (a) $3H_2O + P_4 + 3OH^- \longrightarrow PH_3 + 3H_2PO_2^-$

(b) $12H^+ + 12Cu + 12Cl^- + As_4O_6 \longrightarrow 12CuCl + 4As + 6H_2O$

(c) $9H_2O + 5IPO_4 \longrightarrow I_2 + 3IO_3^- + 5H_2PO_4 + 8H^+$

(d) $3NO_2 + H_2O \longrightarrow 2NO_3^- + NO + 2H^+$

(e) $12OH^- + 6Br_2 \longrightarrow 10Br^- + 2BrO_3^- + 6H_2O$

(f) $4HSO_2NH_2 + 6NO_3^- \longrightarrow 4SO_4^{2-} + 2H^+ + 5N_2O + 5H_2O$

(g) $4H^+ + 2ClO_3^- + 2Cl^- \longrightarrow 2ClO_2 + Cl_2 + 2H_2O$

(h) $2OH^- + 2ClO_2 \longrightarrow ClO_2^- + ClO_3^- + H_2O$

(i) $6OH^- + 3Se \longrightarrow 2Se^{2-} + SeO_3^{2-} + 3H_2O$

(j) $6H_2O + 12ICl \longrightarrow 5I_2 + 2IO_3^- + 12Cl^- + 12H^+$

(k) $4OH^- + 2FNO_3 \longrightarrow O_2 + 2F^- + 2NO_3^- + 2H_2O$

(1) $2H_2O + 4Fe(OH)_2 + O_2 \longrightarrow 4Fe(OH)_3$

6.39   (a) $Zn + 2H_2O + 2OH^- \longrightarrow Zn(OH)_4^{2-} + H_2$

   (b) $2CrO_2^- + 3HO_2^- \longrightarrow 2CrO_4^{2-} + H_2O + OH^-$

6.40   Molarity is the number of moles of solute per liter of solution. Normality is the number of equivalents of solute per liter of solution. The equivalence point occurs when an equal number of equivalents of the reactants have been combined.

6.41   Dissolve 112 g (1.50 mol) of solid KCl in enough water to make a total volume of one liter of solution.

6.42   number of grams of solute per 100 g of solution

6.43   number of grams of solute per million (10^6) grams of solution

6.44   number of grams of solute per billion (10^9) grams of solution

6.45   For acids and bases, an equivalent is the amount of substance that supplies or reacts with one mole of H^+. For redox, an equivalent is the amount of substance that gains or loses one mole of electrons.

6.46   The number of equivalents of A that react is exactly equal to the number of equivalents of B that react.

6.47 $KMnO_4$ is a convenient titrant because it is a powerful oxidizing agent and because of its color. MnO_4^- is purple, whereas Mn^{2+} is almost colorless.

6.48   It doesn't change.

6.49   Add the concentrated reagent to water.

CHAPTER 7

GASES

Rationale

Chapters 7 and 8 treat the three physical states of matter and transitions among them. This begins with gases in Chapter 7. Emphasis is on the gas laws and their manipulation, and on the way that the kinetic theory accounts for the behavior of gas. Finally, the nonideal behavior of gases is shown to lead us to an estimate of molecular size and the strengths of intermolecular attractions.

Although the contents of Chapters 7 and 8 work well in sequence, you can treat the gas laws at an earlier point in the course if your laboratory sequence makes this desirable.

Objectives

At the conclusion of this chapter the students should be able to:

Describe the concept of pressure, the workings of a barometer, as well as open-end and closed-end manometers.

Define the pascal, the standard atmosphere and the torr - the units of pressure.

Illustrate Boyle's, Charles', Dalton's and Gay-Lussac's laws.

Define absolute temperature, Kelvin or absolute scale of temperature, standard temperature and pressure, partial pressure, and vapor pressure.

Apply Boyle's, Charles', Dalton's and Gay-Lussac's laws in problem solving.

Use the ideal gas law in calculations.

Use Graham's law of effusion to predict relative rates of effusion and diffusion of gases.

State the postulates of the kinetic molecular theory.

Account qualitatively for Boyle's, Charles', Dalton's, Gay-Lussac's, Avogadro's and Graham's laws by applying the kinetic molecular theory.

Sketch the kinetic energy distribution in gases at low and high temperatures.

Specify assumptions and limitations of ideal gas behavior.

Explain the effects of molecular size and intermolecular attractions on the behavior of real gases.

Answers to Questions

7.1   Pressure is force per unit area. As long as there is a space above the mercury column the pressures acting along the reference level will be the same regardless of the size and length of the tube. If the diameter of the tube is doubled, there will be twice the weight of Hg acting over twice the area. The ratio of F/A remains unchanged.

7.2   The SI unit of pressure is the pascal; 1 atm = 101,325 Pa

7.3   Sketch would be similar to that shown in Figure 7.3(c). The difference in the height of mercury in the two arms would be 25 mm.

7.4   It is not necessary to measure the atmospheric pressure in order to use it.

7.5   It is dense, it is a liquid over a large temperature range, and it has a very low vapor pressure.

7.6   Boyle's law states that at a constant temperature, the volume occupied by a fixed quantity of gas is inversely proportional to the applied pressure. All gases do not always obey Boyle's law. A gas that does would be called an ideal gas.

7.7   Charles' law states that at a constant pressure, the volume of a given quantity of a gas is directly proportional to its absolute temperature. A temperature of -273°C represents that temperature below which gases would have a negative volume. That is impossible.

7.8   During the trip the tires become warm and the air pressure in them rises (Gay-Lussac's law).

7.9   Heating the can causes the pressure of the gas inside to rise. This causes the can to explode.

7.10   Cooling the gas causes it to contract in volume. This means that a given volume will contain more oxygen at the lower pressure.

7.11   The hot air in the balloon has a lower density than the air surrounding the balloon.

7.12   STP stands for standard temperature and pressure, 0°C (273 K) and one standard atmosphere (760 torr). It is a reference set of conditions.

7.13   Dalton's law states that the total pressure exerted by a mixture of gases is equal to the sum of the partial pressures of each gas in the mixture. Partial pressures cannot be measured after the gases are mixed. However, given sufficient data, partial pressures can be deduced.

7.14   Gay-Lussac's law states that at a constant volume, the pressure of a given quantity of gas is directly proportional to the absolute temperature. Avogadro's principle states that under conditions of constant temperature and pressure, equal volumes of gas contain equal numbers of molecules.

7.15   atm liter^2 K^2/mol

7.16   The units used for pressure and volume determine which R you should use.

7.17   Graham's law states that under identical conditions of temperature and pressure the rate of effusion of gases is inversely proportional to the square root of their densities.

7.18   A gas is composed of a large number of infinitesimally small particles that are in rapid random motion, and the average kinetic energy of the gas is directly proportional to the absolute temperature.

7.19   Pressure arises from the impacts of the molecules of the gas with the walls of the container.

7.20   At the same temperature, their average kinetic energies are the same. Since KE = $\frac{1}{2}mv^2$, if the masses of the molecules of one is less than that of the other, the average velocity must be larger, so the product, $\frac{1}{2}mv^2$, can be the same for both.

7.21   Raising temperature increases the average velocity of the molecules and the gas should diffuse more rapidly.

7.22   More molecules are forced into a given volume. This means that each unit area of wall surface has more molecules above it, so there are more molecule-wall collisions per second.

7.23   Raising the temperature tends to increase the pressure because the molecules hit the walls with more force at the higher temperature.

7.24   The transfer of heat energy is from the warm object to the cool one. As the warm object loses heat its temperature decreases as well as the average K.E. of its molecules. The cooler object receiving heat energy begins to warm, increasing its temperature and the average kinetic energy of its molecules. This process will continue until the average kinetic energies of the molecules in both objects become equal.

7.25   See Figure 7.12. The curves are not symmetrical because molecules have a lower limit of speed (zero) but virtually no upper limit.

7.26   Gases that do not obey Boyle's law are said to be nonideal. This is most evident at high pressures and low temperatures.

7.27   As gases expand, the average distance of separation of the molecules increases. Since real molecules in a gas attract each other somewhat, moving the molecules further apart requires an increase in potential energy at the expense of kinetic energy. Thus the average kinetic energy of the molecules decreases, which leads to a decrease in the temperature of the gas.

7.28   The a is the constant that corrects for the intermolecular attractive forces and b is the constant that corrects for the excluded volume of the molecules.

CHAPTER 8

STATES OF MATTER AND INTERMOLECULAR FORCES

Rationale

As its title suggests, the theme of this chapter is the influence that inter-molecular attractive forces have on the properties of the states of matter, with a particular emphasis on liquids and solids. We begin by describing some of the kinds of properties that are affected, and then follow this by a discussion of various types of intermolecular forces. The concept of quantitative measures of energy changes, first introduced in Chapter 4, is examined here again in discussions of heats of vaporization, fusion, and sublimation. In the discussion of solids, the lattice concept is introduced. A major point here is that a single lattice type can be used to describe enormous numbers of different crystal structures.

Objectives

On completion of this chapter, students should be able to:

Explain why the properties of gases are nearly independent of their chemical composition and why the properties of liquids and solids depend so heavily on their chemical makeup.

Define compressibility, diffusion, surface tension, and evaporation, and be able to compare these properties for liquids and solids.

Explain which properties of liquids and solids depend primarily on the closeness of packing and which depend principally on the strengths of intermolecular attractions.

Define: molar heat of vaporization, vapor pressure, critical temperature and critical pressure, boiling point, freezing point, molar heat of fusion and molar heat of crystallization, sublimation, molar heat of sublimation, supercooling, triple point, and phase diagram.

Describe dipole-dipole forces, hydrogen bonding, and London forces.

Describe and explain the effects of intermolecular attractive forces on vapor pressure, molar heat of fusion, and boiling point.

Apply Le Châtelier's principle to phase changes.

Describe the properties of crystalline solids and compare them to amorphous solids.

Describe X-ray diffraction and how it is useful in the study of the structures of solids.

Explain what a lattice is and describe the properties of the three cubic unit cells.

Describe the four crystal types and give examples of each.

Describe the properties of the different types of liquid crystals.

Sketch the general shape of a heating curve, a cooling curve, and a phase diagram.

Interpret a phase diagram by identifying equilibrium lines and regions where only one phase can exist.

Answers to Questions

8.1

	density	rate of diffusion	compressibility	ability to flow
Solid	high	low	very small	poor
Liquid	medium	medium	very small	good
Gas	low	high	high	good

8.2 density, rate of diffusion, compressibility

8.3 The distance traveled by a molecule between collisions in a liquid is very short compared to a gas. It therefore takes a given molecule more time to move a given distance in the liquid.

8.4 It is a measure of the energy needed to increase the surface area of a liquid. They minimize their surface area and therefore lower their energy.

8.5   Surface tension keeps the surface area of the liquid from expanding, which would happen if the water overflowed.

8.6

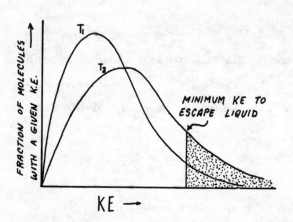

$T_1 < T_2$ It is clear from this diagram that more molecules possess the minimum K.E. for the evaporation at T_2. Therefore, the liquid at T_2 will evaporate faster.

8.7   As the liquid evaporates, the molecules possessing the higher K.E. are leaving, which leads to a lower average K.E. A lower average K.E. means a lower temperature.

8.8   On a dry day water evaporates rapidly because there is little water vapor in the air. The rate of return of H_2O to the clothes is slow compared to the rate of evaporation. When the wind is blowing the air immediately surrounding clothes does not have a chance to saturate. Thus evaporation can continue at a rapid rate.

8.9   It sublimes at the low pressures found at high altitudes.

8.10   A greater fraction of molecules in the warm water have enough energy to escape the surface.

8.11   Sublimation

8.12   Increasing the surface area increases the overall rate of evaporation.

8.13   methyl alcohol

8.14   Attractions between the positive end of one dipole and the negative end of another. They are weaker in a gas because the molecules are further apart.

8.15   Hydrogen bonds are extra strong dipole-dipole attractions. They are most important for N—H, O—H, and F—H bonds because N, O, and F are very small and very electronegative.

8.16 Instantaneous dipole-induced dipole attractions.

8.17

	dipole-dipole	hydrogen bonding	London forces
HCl	x		x
Ar			x
CH_4			x
HF		x	x
NO	x		x
CO_2			x
H_2S	x		x
SO_2	x		x

8.18 Hydrogen bonding forces water molecules into a tetrahedral arrangement about each other which leads to a more "open," less dense structure for ice than for the liquid.

8.19 Surface tension, vapor pressure, ΔH_{vap}, boiling point, freezing point, and ΔH_{fus}.

8.20 Substance X should have the higher boiling point. Substance Y would be the least likely to hydrogen bond.

8.21 ΔH_{vap} and boiling points increase from CH_4 to $C_{10}H_{22}$ because of increased London forces in the same direction. Long chain-like molecules are attracted to one another in more places than shorter molecules.

8.22 ΔH_{vap} should increase $PH_3 < AsH_3 < SbH_3$

8.23 ΔH_{vap} should increase $H_2S < H_2Se < H_2Te$

8.24 There are more hydrogen bonds in water than in HF because H_2O has 2 hydrogen atoms whereas HF has only one.

8.25 The gasoline evaporates very rapidly, thereby removing heat faster than the evaporation of water.

8.26 They should increase.

8.27 Polarizability is the ease of distortion of the electron cloud of an atom, molecule or ion. The greater the polarizability, the stronger the London forces.

8.28 Vapor pressure is the pressure exerted by the gaseous molecules of a substance above a liquid in a closed system at equilibrium.

8.29 If the beverage is cold enough, water in the air immediately surrounding the glass will be cooled sufficiently to condense it on the vessel.

8.30 A humidity of 100% implies air saturated with H_2O vapor, which exerts a partial pressure equal to the vapor pressure of water at that temperature. Vapor pressure increases with temperature; therefore, the amount of water in a given volume of saturated air also increases.

8.31 As the warm moist air rises the temperature begins to fall. The decrease in temperature brings about a lowering of the average K.E., rendering more molecules capable of condensing. This increased condensation is in the form of rain.

8.32 The source of energy in a thunderstorm is the heat of vaporization that is liberated when H_2O condenses.

8.33 When steam contacts skin it condenses to H_2O at 100°C. Heat equal to the heat of vaporization is given off to the skin. Then the skin is further burned by the hot water produced during condensation.

8.34 The equilibrium vapor pressure depends on the rate of evaporation of the liquid, which is determined by the fraction of molecules having enough K.E. to escape. If the attractive forces are large, this fraction is small.

8.35 Decreasing the volume of a container will simply force more vapor molecules to condense into the liquid leaving the V.P. constant. See Figure 8.14, p 256.

8.36 (a) Increasing the temperature will cause a shift to the right.
(b) Increases in temperature will cause a shift to the right.

8.37 It will shift the equilibrium to the left.

8.38 The critical temperature is that temperature above which the substance can no longer exist as a liquid regardless of the pressure. The critical pressure is that pressure required to liquefy a substance at its critical temperature.

8.39 It disappears.

8.40 Below its critical temperature, -82.1°C (Table 8.2, p 259).

8.41   On a cool day the CO_2 is below its critical temperature (31°C, or 88°F) and a separate liquid phase exists.

8.42   The water is able to evaporate more quickly and this cools the coffee.

8.43   That temperature at which the vapor pressure of a liquid is equal to the atmospheric pressure is known as the boiling point of the liquid. The boiling point of a liquid at one standard atmosphere, 760 torr, is referred to as its standard or normal boiling point.

8.44   88°C

8.45   75°C

8.46   The stronger the intermolecular attractive forces, the more difficult it is for molecules to break away from the liquid and enter the gaseous state.

8.47   See Figure 8.17.

8.48   From solid $\longrightarrow$ liquid only small changes in separation take place and only small increases in P.E. occur. When a liquid evaporates, large changes in intermolecular distances occur with correspondingly larger P.E. changes.

8.49   Crystalline solids possess faces that intersect each other at characteristic interfacial angles.

8.50   An amorphous solid has particles arranged in a chaotic fashion. Amorphous solids do not have the characteristic faces and angles of crystals.

8.51   A lattice is a regular or repetitive pattern of points or particles. A unit cell is the smallest grouping of particles that is repeated throughout the solid that can generate the entire lattice. We can create an infinite number of chemical structures by simply varying the chemical environment about each point in a lattice.

8.52   With the use of the Bragg equation, $2d \sin \theta = n\lambda$, we can find the spacing between the successive layers: d is the spacing between successive layers, θ is the angle at which X rays enter and leave, λ is the wavelengths of the X rays and n is an integer.

8.53   The quantities a, b & c, corresponding to the edge lengths of the cell and α, β & γ, corresponding to the angles at which the edges intersect one another, describe a particular lattice.

8.54   See Figure 8.26(a), (b), and (c).

8.55   See Pages 268-269. The rock salt structure belongs to the face-centered cubic lattice. There are four positive ions and four negative ions per unit cell. A salt like K_2S cannot crystallize in the rock salt structure, because it does not have the necessary 1:1 cation to anion ratio.

8.56   Please see Table 8.5, Page 272.

8.57   (a) molecular (b) molecular (c) metallic (d) ionic (e) covalent (f) ionic
        (b) molecular

8.58   (a) molecular (b) ionic (c) ionic (d) metallic (e) covalent (f) molecular
        (g) ionic

8.59   molecular

8.60   covalent

8.61   molecular

8.62   molecular

8.63   ionic

8.64   Nematic, smectic, and cholesteric. They differ in the way the rodlike molecules are packed.

8.65   The color is produced by constructive and destructive interference of light waves reflected from layers in the cholesteric liquid crystal. The particular color observed depends on the distance between the layers, which changes with temperature.

8.66

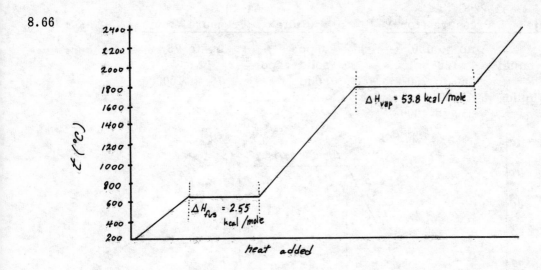

8.67 The triple point of I_2 occurs at a relatively high temperature and at a pressure above atmospheric pressure.

8.68 Camphor, naphthalene (mothballs).

8.69

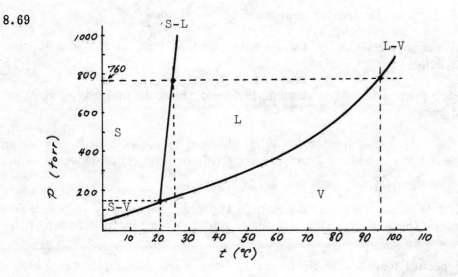

8.70

At 22°C	
State	Pressure (torr)
vapor	up to 160
vapor-liquid	160
liquid	160 to 250
solid-liquid	250
solid	250 to 1000

At 10°C	
State	Pressure (torr)
vapor	up to 75
solid-vapor	75
solid	75 to 1000

8.71

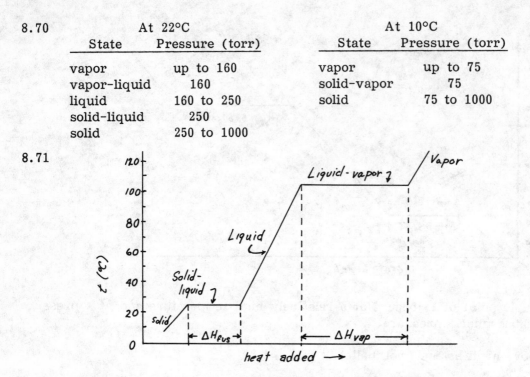

8.72   The density of the solid is greater than that of the liquid.

8.73   The temperature cannot rise above the melting point as long as there is liquid in contact with solid.

8.74   Glass is an amorphous solid. The disorientation of the molecules gives rise to a melting point range.

8.75   When water is cooled, it reaches 32°F (0°C) where it is present as liquid water until further heat is extracted. Then freezing will begin. The answer to the question is yes.

8.76   All the heat that is added is used to increase the potential energy of the molecules. Therefore, the average kinetic energy and the temperature remain constant.

8.77   Increases in pressure lead to the production of the more dense phase. Therefore, increases in pressure will cause the production of more liquid in the case of water. Increases in pressure will cause the production of more solid in the case of CO_2.

8.78   Solid, vapor, solid, vapor, solid, liquid.

CHAPTER 9

THE PERIODIC TABLE REVISITED

Rationale

Now that students have been exposed to basic concepts of chemical bonding and the behavior of the states of matter, a more meaningful discussion of the properties of the elements and their relationship to position in the periodic table can be undertaken. The thrust of this chapter is trends in the properties of the elements. This will serve as background for future discussions of principles and also as a framework about which additional topics in descriptive chemistry can be woven at a later time.

Objectives

When students have completed this chapter, they should be able to:

Mark off regions in the periodic table where metals, nonmetals, and metalloids are to be found.

Describe the physical properties of metals: luster, malleability, ductility, and electrical and thermal conductivity.

Account for the trends in melting point and hardness going from Group IA to IIA.

Identify regions in the periodic table where metals with high or low melting points are to be found.

Identify regions in the periodic table where metals with large or small tendencies to be oxidized are to be found.

Write equations for the reactions of the active metals with water.

Write equations for the reaction of metals with acids.

Write equations for the oxidation of copper by dilute and concentrated nitric acid.

Describe the vertical trend in the stabilities of high and low oxidation states of metals in Groups IIIA through VA.

Describe horizontal and vertical trends in metallic character of the elements in the periodic table.

Explain variations in ionic-covalent character of metal-nonmetal bonds using the concept of ionic potential.

Explain the origin of the colors of metal compounds having charge transfer absorption bands.

Describe the physical properties that are common to nonmetals.

Use the band theory of solids qualitatively to explain the differences between conductors, nonconductors, and semiconductors.

Describe the molecular structures of the common nonmetals and metalloids.

Discuss variations in the strengths of oxoacids and binary acids of the nonmetals.

Answers to Questions

9.1   Because it can be used to correlate large amounts of chemical facts.

9.2   Check answer using Figure 9.1.

9.3   They are nearly equal.

9.4   (a)   any element in an A-group
       (b)   any element in Group IA
       (c)   any element in Group IIA
       (d)   any element in the lanthanides or actinides
       (e)   any element in Group 0

9.5   Malleability - ability to be hammered into sheets
       Ductility - ability to be drawn into wire
       Luster - characteristic sheen that a metal has

Electrical and thermal conductivity - ability to conduct heat and electricity

9.6   Malleability - forging metal parts
       Ductility - making electrical wire
       Luster - mirrors
       Electrical conductivity - electrical wire

9.7   Positive ions at lattice points which are surrounded by a "sea of electrons."

9.8   They haven't enough electrons in their valence shells to form enough bonds to complete their octets.

9.9   Because of the mobile electrons in the metallic lattice.

9.10   Because mobile electrons can serve as carriers of the kinetic energy, transporting it easily through the lattice.

9.11   Ductility

9.12   Malleability

9.13   Tungsten

9.14   They increase toward the center, and then decrease going further to the right.

9.15   In period 6 after the lanthanides, especially in Group VIB and VIIB.

9.16   Because the ions in the Na lattice only have a 1+ charge.

9.17   Calcium

9.18   Positive oxidation states

9.19   (a) Rb   (b) Rb   (c) Na   (d) Ca

9.20   They all react with water to liberate hydrogen.

9.21   Because they are indifferent toward oxidizing agents that attack other metals.

9.22   (a) $Mg + 2HCl \longrightarrow MgCl_2 + H_2$
       (b) $2Al + 6HCl \longrightarrow 2AlCl_3 + 3H_2$

9.23 (a) $2Cr + 6HCl \longrightarrow 2CrCl_3 + 3H_2$

     (b) $Ni + H_2SO_4 \longrightarrow NiSO_4 + H_2$

9.24 For 9.22:

    (a) $Mg + 2H^+ \longrightarrow Mg^{2+} + H_2$

    (b) $2Al + 6H^+ \longrightarrow 2Al^{3+} + 3H_2$

    For 9.23:

    (a) $2Cr + 6H^+ \longrightarrow 2Cr^{3+} + 3H_2$

    (b) $Ni + 2H^+ \longrightarrow Ni^{2+} + H_2$

9.25 Copper, silver, and mercury

9.26 Sn^{4+} is more stable than Pb^{4+}. PbO_2 should be the stronger oxidizing agent.

9.27 More energy is needed to remove electrons than is recovered by bond formation.

9.28 Following the lanthanides in period 6.

9.29 1 part concentrated HNO_3 and 3 parts concentrated HCl, by volume.

9.30 (a) [Ar] $3d^{10}$ (f) [Xe] $4f^{14}5d^8$

    (b) [Ar] $3d^6$ (g) [Kr] $4d^{10}$

    (c) [Ar]   (h) [Ar] $3d^6$

    (d) [Ar] $3d^2$ (i) [Ar] $3d^5$

    (e) [Kr] $4d^{10}$ (j) [Ar]

9.31 Their outer shells have a pair of electrons in an s orbital.

9.32 Ca should react more rapidly because it has a lower ionization energy than Mg and therefore should be more reactive.

9.33 The lower the electronegativity, the more metallic the element. Metallic character decreases left to right and increases from top to bottom. Lithium is more metallic than Be, which is more metallic than B. Carbon is less metallic than Si, which is less metallic than Ge. (In Group IVA, C is a nonmetal and Pb is a metal.)

9.34 (a) Li (b) Al (c) Cs (d) Sn (e) Ga

9.35   (a) $Ca + 2H_2O \longrightarrow Ca(OH)_2 + H_2$

        (b) $2Na + 2H_2O \longrightarrow 2NaOH + H_2$

9.36 Ga_2O_3

9.37 Al_2O_3

9.38   Amphoteric - able to function as either an acid or a base.

$$2Al + 6H^+ \longrightarrow 2Al^{3+} + 3H_2$$

$$2Al + 2OH^- + 2H_2O \longrightarrow 2AlO_2^- + 3H_2$$

$$Be + 2H^+ \longrightarrow Be^{2+} + H_2$$

$$Be + 2OH^- \longrightarrow BeO_2^{2-} + H_2$$

9.39 $Al_2O_3 + 2OH^- \longrightarrow 2AlO_2^- + H_2O$

9.40 ϕ = charge/ion radius

9.41   (a) $GeCl_4$ (b) Bi_2O_5 (c) PbS (d) Li_2S (e) MgS

9.42   (a) SnO (b) $AlCl_3$ (c) BeF_2 (d) PbS (e) SnS

9.43   Charge transfer from anion to cation which absorbs photons in the visible portion of the spectrum.

9.44   (a) Ag_2S (b) $CuBr$ (c) SnS_2 (d) Al_2S_3

9.45   Red

9.46 H_2, N_2, O_2, F_2, Cl_2, Group 0

9.47   Bromine and mercury

9.48   Their valence bands are full and there is a large separation between the valence and conduction band.

9.49   In conductors, the valence band is partially empty or overlaps an empty conduction band. In a nonconductor, the valence band is filled and separated from the conduction band by a large energy gap. In a semiconductor, the gap between valence band and conduction band is small.

9.50   An n-type semiconductor results when one metalloid is "poisoned" with small amounts of another that contains an extra electron. These "extra" electrons are free to move when a voltage is applied.
A p-type semiconductor results when one metalloid is doped with small amounts of another that contains less electrons. When a voltage is applied, electrons flow to fill the deficit, creating others that need to be filled.

9.51   (a) Ga, In   (b) As, Sb

9.52   Light energy that is absorbed releases electrons trapped in p-layer of a p-n junction and permits these electrons to flow back into the n-layer. In effect, this institutes a current flow.

9.53   Thermal energy raises electrons from valence band to conduction band. The higher the temperature, the greater the electron population in the conduction band.

9.54   One of the dominant factors in determining the complexity of the molecular structure is the ability of the second period nonmetals to enter into multiple bonding and the tendency of the heavier elements to prefer single bonds.

9.55   Their large size prevents them from approaching each other closely enough to give effective sideways overlap of p orbitals, which is necessary for π-bond formation.

9.56   Graphite and diamond. Graphite consists of layers of carbon atoms in which in a given layer the C atoms are arranged in joined 6-membered rings. The delocalized π-cloud of each layer permits electrical conductivity, and the layers slip over each other easily. In diamond each C atom is joined to four others by single covalent bonds which gives a strong interlocking network. Graphite is a black solid. Diamond crystals, of course, are brilliantly clear.

9.57 N_2 contains a triple bond between N atoms. P_4 takes the shape of a tetrahedron with each P singly bonded to three others (see Figure 9.22). The P—P—P bond angle is approximately 60°. If the bonding is of pure p-p type, the bond angle should be 90°. Thus the bonds in white phosphorus are quite strained, and it should be very reactive.

9.58   White phosphorus consists of 4 P atoms at the corners of a tetrahedron whereas black phosphorus has a layer structure in which each P atom in a layer is singly bonded to three others.

9.59   The free movement of electrons in the delocalized π-cloud that extends over the graphite sheet accounts for graphite's ability to conduct electricity.

9.60   Icosahedron (see Figure 9.23).

9.61   The gradual conversion of metallic tin to its nonmetallic allotrope, which causes tin objects to deteriorate in appearance.

9.62   (a) $HClO_3$ (b) HNO_3 (c) H_3PO_4 (d) $CHCl_2COOH$ (e) CH_2FCOOH (f) H_2SO_4

9.63   (a) H_2Se (b) HBr (c) PH_3

9.64   The additional oxygen in HNO_3 draws more electron density from the nitrogen which gives it a greater partial positive charge. The nitrogen in turn draws electron density from the N—OH bond. This in turn draws the electron density from the O—H bond making HNO_3 a stronger acid than HNO_2.

9.65 Cl^- is a larger atom than F^- so that the HCl bond is much weaker than the HF bond, thus making it more acidic.

9.66   HCl is much more polar than H_2S.

9.67 CH_3SH

CHAPTER 10

PROPERTIES OF SOLUTIONS

Rationale

This second chapter on solutions, which focuses on the physical properties of solutions as they are affected by the concentration of the solute, follows naturally from the discussions in Chapters 7 and 8 concerning the physical properties of pure substances. The groundwork is laid by specifying the various ways that solution concentration may be defined. For completeness, molarity and normality (previously discussed in Chapter 6) are defined again.

The solution process is examined here in detail in terms of the energy changes that occur. This provides a basis for the definition of an ideal solution later in the chapter. Solubility as an equilibrium process provides an opportunity for the application of Le Châtelier's principle (which was introduced in Chapter 8).

Raoult's law provides an introduction to colligative properties which are discussed further in terms of freezing point depression, boiling point elevation, and osmotic pressure. Raoult's law also serves as an introduction to fractional distillation.

Objectives

Upon completion of this chapter, students should be able to:

Define the concentration units: mole fraction, weight fraction, weight percent, molality, molarity, normality.

Carry out conversions among the various concentration units.

Analyze the solution process in terms of the factors that contribute to the heat of solution.

Define: hydration, solvation, heat of solution, ideal solutions, lattice energy, hydration energy, fractional crystallization, Henry's law, and Raoult's law.

Illustrate graphically: exothermic and endothermic solution processes; positive and negative deviations from Raoult's law; a boiling point diagram for a two-component mixture, with and without an azeotrope.

Relate Le Châtelier's principle to solubility versus temperature and to the effect of pressure on the solubility of a gas.

Define: fractional distillation, tie line, azeotrope, colligative properties, osmosis and osmotic pressure.

Use colligative properties to determine molecular weights.

Answers to Questions

10.1   <u>Substitutional solid solution</u> - particles of solute replace the "solvent" in lattice. <u>Interstitial solid solution</u> - solute particles fit into spaces between "solute" particles in the host lattice.

10.2   Mole fraction = moles solute/total moles of all species in the solution. Mole percent = 100 x mole fraction. Weight fraction = weight of a particular solute divided by the total weights of all components. Weight percent = 100 x weight fraction. Molarity = moles solute/volume of solution (in liters). Molality = moles solute/ kilograms solvent.

10.3   They are ratios (fractions).

10.4   Substances that exhibit similar intermolecular attractive forces tend to be soluble in one another.

10.5   The salt increases the effective polarity of the solvent and therefore decreases the solubility of solutes that are not as polar.

10.6   Methyl alcohol is polar enough to dissolve water and at the same time non-polar enough to be dissolved into the gasoline.

10.7   The ion is surrounded by water molecules.

10.8   In an ideal gas there are no intermolecular attractions. An ideal solution is one in which the solute-solute, solute-solvent and solvent-solvent interactions are all the same (but they are not zero).

10.9   The heat of solution is equal to the difference between the lattice energy and the hydration energy. If the lattice energy is greater, a net input of energy is required, i.e. the solution process is endothermic.

10.10 (a) Na^+ (b) F^- (c) Ca^{2+} (d) Fe^{3+} (e) S^{2-}

10.11 KI would become more soluble with an increase in temperature. Addition of heat favors an endothermic process.

10.12 The dominant energy effect when gases dissolve in a liquid is caused by the solvation of the gas molecules, which is exothermic.

10.13 In fractional crystallization, an impure product is dissolved in a small amount of hot solvent, which is then cooled. As the solution cools, the pure product separates from the mixture, leaving the impurities behind.

10.14 KBr @ 70°C, KNO_3 will crystallize first @ 20°C.

10.15 Increasing the pressure will increase the rate at which molecules leave the gas and enter the solution. This will continue until equilibrium is reestablished, at which time the concentration of the solute in the solution will have increased.

10.16 Pressure only has an appreciable effect on equilibria where sizable volume changes occur. When a liquid or solid dissolves in a liquid only very small changes in volume occur.

10.17 The vapor pressure of a solvent depends on the fraction of the total number of molecules at the surface of the solution that are solvent molecules; i.e. the mole fraction of the solvent.

10.18 When the vapor pressure of a mixture is greater than that predicted, it is said to exhibit a positive deviation from Raoult's law; conversely, when a solution gives a lower vapor than we would expect from Raoult's law, it is said to show a negative deviation.

10.19 ΔH_{sol} for positive deviations is endothermic, whereas ΔH_{sol} for negative deviations is exothermic.

10.20 When a mixture of two liquids is boiled, the vapor is always richer in the more volatile component. Successive condensations and boilings produce fractions ever richer in the more volatile component.

10.21 Approximately three times.

10.22

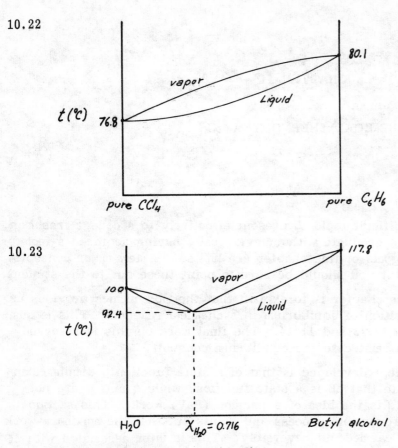

10.23

Positive deviation from ideality

10.24 In the presence of a solute the rate of freezing at a particular temperature is decreased because fewer solvent particles are in contact with the solid. The rate of melting, however, is the same since no solute is incorporated in the solid solvent. The temperature must be lowered to reestablish equilibrium where the solvent freezes faster from the solution and the solvent melts more slowly from the solid.

10.25 $MgSO_4$. The greater the degree of attraction between ions, the less it dissociates.

10.26 A nonelectrolyte that has formed a dimer might give an $\underline{i}$ factor less than one.

10.27 $Al_2(SO_4)_3$

10.28 KCl, i factor = 2; $NiCl_2$, i factor = 3; $Al_2(SO_4)_3$, i factor = 5.

CHAPTER 11

CHEMICAL THERMODYNAMICS

Rationale

Thermodynamics is a difficult topic to present effectively to a college freshman. In fact, many chemistry majors graduate without ever really having acquired a feel for the subject. Therefore, the goals of this chapter are limited. A few major concepts are developed and at some point you should be sure to point these out to the student.

The ultimate goal of the chapter is to explain that thermodynamics provides us with information about the position of equilibrium in a chemical reaction. This is summarized graphically in Figures 11.13 and 11.14. The final word on this is reserved until Chapter 13 where $\Delta G°$ is related to the equilibrium constant.

Another important concept developed is that of a state function. Simple calculations in Section 11.2 illustrate that ΔE is a state function, while q and w are not. Also introduced in Section 11.2 is the idea of expansion ($P\Delta V$) work. This section also introduces the idea of a reversible process and points out that the maximum work is obtained when a process is carried out reversibly. This is later associated with ΔG and serves as a basis for relating thermodynamic efficiency to how close a process comes to actually being reversible.

ΔH is introduced as the heat of reaction at constant pressure. This is followed by a discussion of thermochemistry. (Note: Much of this could be presented earlier in the course, if you desire.)

You will note that calculations in this chapter are given sometimes in joules and at other times in calories; tables contain entries in both joules and calories. Even after the sciences have switched entirely to the SI units, scientists will still have to know both because the past literature has energies expressed in calories. The approach taken in the text with regard to energy units reflects the desirability of familiarizing students with both joules and calories.

Sections 11.7 and 11.8 introduce the notion of entropy and relate changes in energy and entropy to reaction spontaneity. This is followed by a "derivation" of the equation, $\Delta G = \Delta H - T\Delta S$, which shows how these factors (ΔH, ΔS) are related quantitatively.

Depending on the ability of your students, you may wish to give a treatment of thermodynamics that is more qualitative than quantitative. Alternate paths through the chapter that would allow you to omit some of the calculations are given below.

Section 11.1   Introduce appropriate terminology.

Section 11.2   Introduce internal energy and $\Delta E = q - w$. Point out that for expansion $w = P\Delta V$. Explain that ΔE is a state function, q and w are not.

Section 11.3   Explain calorimeter calculation, $\Delta E = q$ at constant V. Define enthalpy, $\Delta H = q$ at constant P. You can say that in most instances $\Delta H \approx \Delta E$ (omit Example 11.3).

Section 11.4   Hess's law and heats of reaction. Cover this thoroughly.
and 11.5

Section 11.6   Optional. This section shows why ΔH calculations are useful, but the material is also somewhat complex. If time is short, this could be omitted.

Section 11.7   Explain how an energy decrease and an increase in statistical probabil-
and 11.8   ity (randomness) are two factors that favor spontaneity. Tie entropy into statistical probability (you can omit $\Delta S = q/T$).

Section 11.9   Go directly to $\Delta G = \Delta H - T\Delta S$ which relates ΔH, ΔS. Point out that a spontaneous process occurs with a decrease in G ($\Delta G < 0$).

Section 11.10   This is worth mention because idea of efficiency of energy conversion is introduced.

Section 11.11   These are simply Hess's law-type calculations.

Section 11.12   This section brings into focus one of the principal goals of the chapter - to be able to use thermodynamics to predict whether or not a reaction can be expected to be observed either in the laboratory or elsewhere.

A still further abrided treatment could begin with Section 11.3 on heats of reaction after introducing some terminology from Section 11.1 (e.g., state function, specific heat, heat capacity, etc.).

Objectives

Upon completion of this chapter, students should be able to:

Define: system, surroundings, adiabatic process, isothermal process, state, state function, equation of state, heat capacity, and the first law of thermo-dynamics.

Use the concepts described in the first law of thermodynamics to calculate (for simple systems): the change in internal energy, the amount of heat added to or removed from a system, or the work done by or on a system.

Distinguish between reversible and irreversible processes.

Define enthalpy, thermochemical equation, Hess's law of heat summation, heat of formation, atomization energy, spontaneity, and entropy.

Apply Hess's law to thermochemical equations and calculate the heat of an overall reaction.

Use heats of formation to calculate the heat of an overall reaction.

Compute the heat of formation of some simple molecules from bond energies.

Define the second law of thermodynamics, Gibbs free energy, useful work, and the third law of thermodynamics.

Calculate ΔS° from tabulated S°.

Calculate ΔG° using ΔG_f°.

Use a calculated ΔG° to decide whether a reaction is feasible.

Describe how the Gibbs free energy changes during a chemical reaction and relate the free energy to equilibrium.

Relate qualitatively the value of ΔG° to the position of equilibrium in a chemical reaction.

Answers to Questions

11.1   Whether the reaction is spontaneous or not, and how fast the reaction occurs.

11.2   Thermo - heat, dynamics - movement or change

11.3   System & Surroundings - By system we mean that particular portion of the universe upon which we wish to focus our attention. Everything else we call the surroundings.
Isothermal change - change occurring at constant temperature.
Adiabatic change - a change that occurs without heat transfer between system and surroundings.

State function - a quantity whose value for a system in a particular state is independent of the system's prior history.

Heat capacity - heat needed to raise the temperature of the system by 1°C.

Molar heat capacity - heat needed to raise the temperature of 1 mol of a substance by 1°C.

Specific heat - heat needed to raise the temperature of one gram of a substance by 1°C.

11.4 $\Delta E = q - w$. The ΔE is the difference between the heat that is added to the system as it passes from the initial to the final state and the work done by the system upon its surroundings. E represents all of the energies, KE + PE, whereas ΔE represents $E_{final} - E_{initial}$. ΔE, therefore, is only dependent on the energy the molecules have finally minus what they had initially, regardless of the path.

11.5   Because of the very nature of E, which represents the KE as well as the PE. For instance, because there is no stationary reference point in the universe, we cannot measure absolute velocity; therefore, we cannot know KE.

11.6   (a)   There would be no change in temperature. There are no attractive forces between the molecules in an ideal gas, so there would be no change in PE when the molecules move further apart.

   (b)   Since gases cool on expansion, the average kinetic energy of the molecules must decrease. Therefore, ΔE is positive, q is positive, w = 0. Heat would have to be supplied to keep the temperature of the system constant (isothermal).

11.7   Because during an isothermal expansion or compression there is no change in the KE or PE of the material. KE remains constant because T remains constant. PE is zero because in an ideal gas there are no intermolecular attractive forces.

11.8   An equation that relates state variables (e.g., P, V, T).

11.9   Pressure - Volume and electrical.

11.10 A spontaneous change is one that will take place by itself, without continued outside aid.

11.11 A reversible process is one that can be made to reverse its direction by an infinitesimal change in pressure, temperature, or whatever opposes the change.

11.12 Advantage - maximum work is obtained.
       Disadvantage - the change takes forever to occur.

11.13 It can't just disappear. It shows up as an increase in the heat of reaction.

11.14 $P\Delta V$ has the units Joules when P is in pascals and ΔV is in m^3.

11.15 Isothermal: immerse the system in a vat of water kept at a constant temperature by a thermostat.
Adiabatic: keep the system in an insulated container.

11.16 $\Delta E = q_V$ because $P\Delta V = 0$.
$\Delta H = q_P$ because $\Delta H = (q - P\Delta V) + P\Delta V$

11.17 ΔE is the heat of reaction at constant volume, whereas ΔH is the heat of reaction at constant pressure. Most reactions that are of interest to us take place at constant P, not constant V.

11.18 A heavy metal container (in which a reaction can be carried out) that is immersed in a large insulated container filled with water. Its purpose is to measure heats of reaction at constant volume.

11.19 Hess's law of heat summation states that the ΔH of an overall process is merely the sum of all of the enthalpy changes that take place along the way. Conditions of 25°C and one atmosphere pressure are taken to be the standard state of a substance.

11.20 No; only $\Delta H°$ for the second reaction should be labeled ΔH_f^o because only in that reaction is SO_3 formed from its elements.

11.21 The second reaction (because of the heat released when gas is condensed to liquid).

11.22 Reactions that involve only liquids and/or solids. ΔE and ΔH rarely differ by much, even when gases are formed or produced.

11.23 Atomization energy is the sum of all of the bond energies in the molecule.
$H_2O(g) \longrightarrow 2H(g) + O(g)$
$\Delta H = \Delta H_{atomization}$

11.24 The calculated ΔH_f^o is usually from average bond energies.

11.25 In any process there is a natural tendency or drive toward increased randomness because a highly random distribution of particles represents a condition of higher statistical probability than an ordered one. A system's entropy is proportional to its statistical probability.

11.26 In order for Δs to be a state function the value of q must be a state function, and q_{rev} is a state function.

11.27 There is perfect order (zero randomness) in a pure crystalline substance at absolute zero. A mixture would have a positive entropy at 0 K because of the random distribution of particles throughout the mixture.

11.28 ΔG must be negative which means ΔH must be negative and ΔS must be positive.

11.29 During any spontaneous change, the entropy of the universe increases.

11.30 (a) negative (b) positive (c) positive (d) negative (e) negative
(f) negative (g) positive

11.31 $\Delta G°$ determines the position of equilibrium between reactants and products. Whether we start with pure reactants or pure products, some reaction will occur (accompanied by a free energy decrease) until equilibrium is reached.

11.32 The sign and magnitude of $\Delta G°$ tells us where the reaction is going, in the sense that it relates to the position of equilibrium.

11.33 There is none.

11.34 See Figure 11.14.

11.35 Between 10,000 and 20,000 atm

11.36 Refer to Figure 11.13. Once a minimum free energy has been achieved by the system, the composition of the system can no longer change, since such a change involves going "uphill" on the free energy curve. At the minimum, both reactants and products possess the same free energy and therefore, $\Delta G = 0$.

11.37 $\Delta G = \Delta H - T\Delta S$

Note temperature in the equation.

CHAPTER 12

CHEMICAL KINETICS

Rationale

Thermodynamics and kinetics are presented in sequence because thermodynamic feasibility and reaction rate dictate whether or not the products of a given reaction will be observed. Section 12.1 begins by explaining what reaction rate means, and how it is expressed and measured. Then, in Section 12.2, rate is related to the concentrations of the reactants. A point made repeatedly here is that the exponents in the rate law can only be obtained experimentally. The relationship between concentration and time, and the concept of half-life of a reaction are discussed next in Section 12.3.

Collision theory is developed to account for the mathematical form of the rate law. This is then used as a basis for sorting through alternative proposed reaction mechanisms. The goal of Section 12.5 is to show the student how reaction mechanisms are derived from kinetic studies.

Sections 12.6 to 12.8 examine the importance of molecular orientation and energies during collisions. The calculation of E_a from rate constants at different temperatures is described.

In Section 12.9 it is important to emphasize that a catalyst lowers the activation energy for a reaction by opening up an alternative, low energy path from reactants to products.

The discussion of chain reactions in Section 12.10 introduces the concept of a free radical and describes the types of steps involved in a chain mechanism.

Objectives

At the conclusion of this chapter, students should be able to:

Define: reaction rate, rate constant, rate law, and the order of a reaction.

Compare the relative rates of disappearance of reactants and formation of products in a reaction based on the coefficients in the equation.

Measure reaction rate at a particular time from concentration versus time data.

Calculate $t_{\frac{1}{2}}$ for first- and second-order reactions, and be able to calculate concentration at some time t from the integrated rate laws.

Determine the rate law for a given reaction and calculate the value of its rate constant from initial rate data.

Explain how collision theory allows us to predict the rate law for an elementary process.

Describe reaction mechanisms as a sum of elementary processes.

Define rate-determining step, reaction mechanism, elementary process, bimolecular and termolecular collisions, effective collisions, and energy of activation.

Calculate the energy of activation given rate constants at different temperatures.

Calculate the rate constant at some temperature given E_a and k at another temperature.

Sketch and label potential energy diagrams for reactions.

Define: activated complex, catalyst, chain reaction and free radical.

Apply the concepts in this chapter to describe the behavior of a catalyst.

Identify the types of steps involved in chain reactions.

Answers to Questions

12.1 (a) the nature of reactants and products
      (b) concentration of reacting species
      (c) temperature
      (d) influence of outside agents (catalysts)

12.2 The smaller the particle size (i.e., the larger the surface area), the faster the reaction.

12.3 Reaction rate = the speed at which reactants are consumed or the products are formed. It is the ratio of the change in concentration to the change in time. units = mol liter^{-1} s^{-1}

12.4 (a)
$$\text{Rate} = \frac{-\Delta[H_2]}{\Delta t} = \frac{-2\Delta[O_2]}{\Delta t} = \frac{\Delta[H_2O]}{\Delta t}$$

(b)
$$\text{Rate} = \frac{-\Delta[NOCl]}{\Delta t} = \frac{\Delta[NO]}{\Delta t} = \frac{2\Delta[Cl_2]}{\Delta t}$$

(c)
$$\text{Rate} = \frac{-\Delta[NO]}{\Delta t} = \frac{-\Delta[O_3]}{\Delta t} = \frac{\Delta[NO_2]}{\Delta t} = \frac{\Delta[O_2]}{\Delta t}$$

(d)
$$\text{Rate} = \frac{-\Delta[H_2O_2]}{\Delta t} = \frac{-\Delta[H_2]}{\Delta t} = \frac{1}{2}\frac{\Delta[H_2O]}{\Delta t}$$

(a) H_2 disappears twice as fast as O_2, and same as H_2O appears.
(b) NOCl disappears at same rate as NO appears and twice as fast as Cl_2 appears.
(c) As NO and O_3 disappear, NO_2 and O_2 appear at the same rate.
(d) H_2O_2 and H_2 disappear half as fast as H_2O appears.

12.5 Methods must be fast, accurate and not interfere with the normal course of the reaction system.

12.6 Some common examples:
(a) combustion of gas (rapid)
(b) explosion of gasoline vapor in auto engine (very fast)
(c) digestion of food (moderately slow)
(d) iron rusting (slow)
(e) decay of leaves (slow)

12.7 An experimentally determined relationship between the rate of reaction and the concentrations of the reactants. Temperature and catalysts affect k.

12.8 The sum of the exponents on the concentrations in the rate law.

12.9 (a) s^{-1} (b) liter mol^{-1} s^{-1} (c) liter^2 mol^{-2} s^{-1}

12.10 It is simply because some substances react rapidly with one another because of their chemical composition, while others react more slowly. In other words, even under conditions of equal concentrations and temperature, different chemical reactions progress at different rates.

12.11 As the concentration of CO varies, the rate at which it is removed remains constant (no concentration dependence); therefore, it appears to be a zero-order reaction.

12.12 (a) first order with respect to A and B, overall order is two.
(b) second order with respect to E, overall order is two.
(c) second order with respect to G and H, overall order is four.

12.13 (a) liter/mol s   (b) liter/mol s   (c) liter^3/mol^3 s

12.14 (a) rate doubled
(b) rate increased fourfold
(c) rate increased eightfold
(d) rate increased sixteenfold
(e) rate increased by a factor of $2^{1/2} = 1.44$
(f) rate decreased by a factor of 4 ($2^{-2} = 1/4$)

12.15 -1; i.e., Rate = $k[A]^{-1}$

12.16 The rate of a reaction is proportional to the number of collisions per second between reacting molecules. As the number of molecules is increased (increase in concentration), the number of collisions is increased by the same factor.

12.17 An overall chemical reaction represents the net chemical change. This does not mean that all the reactants come together simultaneously. To predict the rate law, a reaction mechanism must be known.

12.18 A reaction mechanism is a series of elementary processes that lead to the formation of the products.

12.19 A one-step mechanism would involve the simultaneous collision of six molecules, five of which would have to be O_2. This is very improbable.

12.20 The time required for the concentration of a given reactant to be decreased by a factor of 2.

12.21 It is unaffected by the initial concentration.

12.22 (a) Rate = $k[NO][Br_2]$
(b) $NO + Br_2 \longrightarrow NOBr_2$
Rate = $k[NO]^2[Br_2]$
(c) Step 2 must be rate determining (i.e., the slow step)

(d) This is a termolecular collision, which is very unlikely for a fairly rapid reaction.

(e) No. A mechanism is only theory, and more information can support it or prove it wrong, but the actual path can never be known with complete certainty.

12.23 The rate law is Rate = $k[NO_2]^2$. CO does not appear in the rate-determining step (slow step), and therefore does not affect the rate.

12.24 $(CH_3)_3CBr \longrightarrow (CH_3)_3C^+ + Br^-$ slow

$(CH_3)_3C^+ + OH^- \longrightarrow (CH_3)_3COH$ fast

for which Rate = $k[(CH_3)_3CBr]$

12.25 (a) $2A + B \longrightarrow C + 2D$
(b) Rate = $k[A]^2$
(c) Rate = $k[A]^2[B]$

12.26 $NO_2 + O_3 \longrightarrow NO_3 + O_2$ slow

$NO_3 + NO_2 \longrightarrow N_2O_5$ fast

12.27 The observed rate of reaction is much smaller (approx. 5×10^{12}) than what is expected if all the collisions were effective. A minimum amount of energy is required to cause a reaction to occur, and the molecules must collide with the proper orientation.

12.28 When two molecules collide, the collision can occur so that products are obtained that are the same as the reactants, or no chemical change occurs at all. Only proper orientation produces effective collisions.

12.29

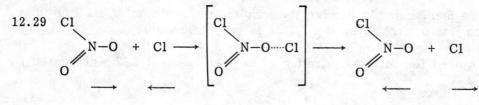

(not an effective collision)

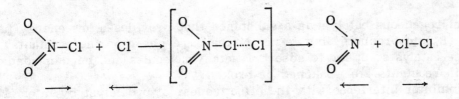

(an effective collision)

12.30 $NO + O_2 \rightleftharpoons NO_3$ fast

$NO_3 + NO \longrightarrow 2NO_2$ slow

12.31 The activation energy represents the energy required to bring the reactants to the point where they can react to form products. It is the minimum energy that must be available in a collision.

12.32 Activation energy is the minimum energy required for reaction. As T is increased more molecules have greater kinetic energy, i.e., a larger fraction of molecules have the minimum energy required for reaction.

12.33

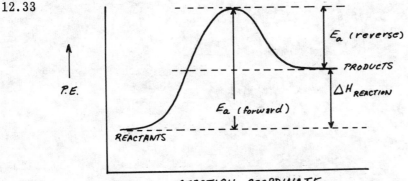

12.34 The activated complex is the intermediate species that is highly unstable, transient and very reactive. The activated complex exists in a transition state between products and reactants. The transition state is on the hump (high potential energy) of the potential energy curve.

12.35 Reactions, including biochemical ones, slow down at low temperatures primarily because a smaller fraction of molecules possesses the required activation energy.

12.36 A homogeneous catalyst is in the same phase as the reactants while a heterogeneous catalyst is in a different phase (e.g., a catalytic surface in contact with reacting gases).

12.37 (a) A heterogeneous catalyst is a substance that provides a low energy path-way (E_a) to products, but is not in the same phase as the reactants. These catalysts appear to adsorb reactant molecules and certain bonds within reactants are weakened or broken.
  (b) An inhibitor interferes with the effectiveness of a catalyst by interfering with adsorption.

12.38 The catalyst lowers the activation energy by giving the reactants a different pathway (mechanism) through a chemical reaction. This path has a lower E_a, thereby increasing the number of effective collisions.

12.39 (a) no effect
  (b) no effect
  (c) lowers activation energy by changing the nature of the transition state.

12.40 In chain reactions a single reactive intermediate is produced (radical) which produces many product molecules before termination. Therefore, products are produced at a rate faster than the initiation step alone.

12.41 (a) initiation step 1
  (b) propagation steps 3,4,5,6
  (c) termination step 2

CHAPTER 13

CHEMICAL EQUILIBRIUM

Rationale

The mass action law is introduced here as an experimentally observable phenomenon and then justified from the point of view of thermodynamics in Section 13.3. The main emphasis in this chapter is on gaseous equilibria, and the possibility of expressing K as either K_c or K_p is brought into the discussion very early. Conversion between K_c and K_p is covered in Section 13.4, followed by extension of the equilibrium discussion to heterogeneous systems.

Section 13.6 provides a thorough discussion of the application of Le Châtelier's principle to chemical equilibria. Students should recall earlier applications to physical equilibria.

The chapter concludes with a number of examples involving equilibrium calculations. To help students assemble the data, encourage them to set up the data tables as shown in the examples. In particular, you should point out that the concentrations that are entered into the K_c expression must always come from the column labeled "Equilibrium Concentrations." Also point out that quantities in the "Change" column always occur in the same ratio as the coefficients in the balanced equation for the equilibrium. In fact, when inserting x's into the "Change" column, the coefficients of x can be the same as the coefficients in the balanced equation. Finally, all reactant changes must have the same algebraic sign. This is opposite in sign to the changes for the products.

Objectives

At the conclusion of this chapter students should be able to:

Define: equilibrium, equilibrium constant, and the law of mass action.

Write the mass action expression for any reaction given a balanced chemical equation.

Make qualitative judgments of the extent of reaction based on the magnitude of K.

Write appropriate equilibrium expressions for heterogeneous equilibria.

Apply Le Châtelier's principle to chemical equilibria.

Perform the following types of calculations: determine thermodynamic equilibrium constants given standard free energy changes, convert between K_c and K_p for any reaction, determine equilibrium constants given equilibrium concentrations, and determine equilibrium concentrations given the equilibrium constant and initial concentrations of reactants.

Answers to Questions

13.1 Products are constantly changing to reactants and reactants are constantly changing to products but the overall effect is no apparent change - equilibrium.

13.2 (a) $\dfrac{[NO]^2}{[N_2][O_2]}$ (b) $\dfrac{[NO_2]^2}{[NO]^2[O_2]}$ (c) $\dfrac{[H_2S]^2}{[H_2]^2[S_2]}$ (d) $\dfrac{[NO_2]^4[O_2]}{[N_2O_5]^2}$

(e) $\dfrac{[POCl_3]^{10}}{[P_4O_{10}][PCl_5]^6}$

13.3 (a) $\dfrac{p_{NO}^2}{p_{N_2}\,p_{O_2}}$ (b) $\dfrac{p_{NO_2}^2}{p_{NO}^2\,p_{O_2}}$ (c) $\dfrac{p_{H_2S}^2}{p_{H_2}^2\,p_{S_2}}$ (d) $\dfrac{p_{NO_2}^4\,p_{O_2}}{p_{N_2O_5}^2}$

(e) $\dfrac{p_{POCl_3}^{10}}{p_{P_4O_{10}}\,p_{PCl_4}^6}$

13.4 (a) $K_p = \dfrac{p_{CH_3OH}}{p_{CO}\,p_{H_2}^2}$, $K_c = \dfrac{[CH_3OH]}{[CO][H_2]^2}$

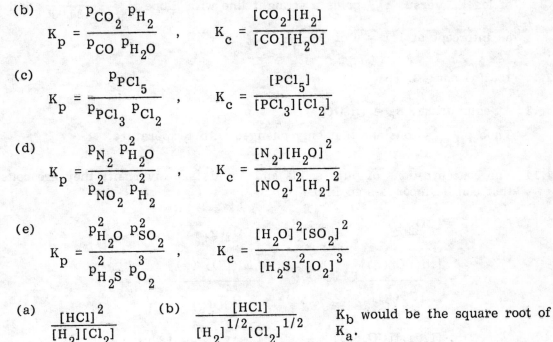

(b)
$$K_p = \frac{p_{CO_2}\, p_{H_2}}{p_{CO}\, p_{H_2O}} \quad , \quad K_c = \frac{[CO_2][H_2]}{[CO][H_2O]}$$

(c)
$$K_p = \frac{p_{PCl_5}}{p_{PCl_3}\, p_{Cl_2}} \quad , \quad K_c = \frac{[PCl_5]}{[PCl_3][Cl_2]}$$

(d)
$$K_p = \frac{p_{N_2}\, p_{H_2O}^2}{p_{NO_2}^2\, p_{H_2}^2} \quad , \quad K_c = \frac{[N_2][H_2O]^2}{[NO_2]^2[H_2]^2}$$

(e)
$$K_p = \frac{p_{H_2O}^2\, p_{SO_2}^2}{p_{H_2S}^2\, p_{O_2}^3} \quad , \quad K_c = \frac{[H_2O]^2[SO_2]^2}{[H_2S]^2[O_2]^3}$$

13.5 (a) $\dfrac{[HCl]^2}{[H_2][Cl_2]}$ (b) $\dfrac{[HCl]}{[H_2]^{1/2}[Cl_2]^{1/2}}$ K_b would be the square root of K_a.

13.6 By convention. This simplifies tabulation of equilibrium constants by removing ambiguity.

13.7 By looking at the size of the equilibrium constant, you can determine whether the reaction favors the forward or reverse reaction. If the number is greater than one, the reaction will tend to proceed far toward completion. If, however, the K is less than one, only small amounts of products will be present at equilibrium.

13.8 From the magnitude of K we can say that the tendency to proceed toward completion increases in the order (b) < (c) < (d) < (a).

13.9 $\Delta G° = 0$

13.10 In Section 13.3 it is stated that for gases, the equilibrium constant calculated from $\Delta G°$ is K_p.

13.11 $-\dfrac{1}{T}\dfrac{\Delta H°}{2.303\ R} + \dfrac{\Delta S°}{2.303\ R} = \log K_p$

$XM + b = Y$, Let $\log K_p$ correspond to Y and $1/T$ correspond to X. A plot of log K_p versus $1/T$ gives a straight line with Slope $M = -\dfrac{\Delta H^\circ}{2.303\ R}$ and an an intercept at $1/T = 0$ of $\dfrac{\Delta S^\circ}{2.303\ R}$.

13.12   13.2(a) and 13.4(b)

13.13   At equilibrium, $K_c = [H_2O(g)]$, $K_p = p_{H_2O(g)}$
Thus p_{H_2O} = constant that only changes with temperature.

13.14   The concentrations of pure solids and liquids are invariant; they are constants that can be incorporated into K_{eq}.

13.15   (a) $K_c = [CO_2(g)]$ $K_p = p_{CO_2(g)}$

(b)
$$K_c = \frac{[Ni(CO)_4(g)]}{[CO(g)]^4} \qquad K_p = \frac{p_{Ni(CO)_4(g)}}{p_{CO(g)}^4}$$

(c)
$$K_c = \frac{[I_2(g)][CO_2(g)]^5}{[CO(g)]^5} \qquad K_p = \frac{p_{I_2(g)}\ p_{CO_2(g)}^5}{p_{CO(g)}^5}$$

(d)
$$K_c = \frac{[CO_2(g)]}{[Ca(HCO_3)_2(aq)]} \qquad K_p = \frac{p_{CO_2}}{[Ca(HCO_3)_2]}$$

(e) $K_c = [Ag^+(aq)][Cl^-(aq)]$

13.16 $PCl_3(g) + Cl_2(g) \rightleftharpoons PCl_5(g)$

(a)   Addition of PCl_3 would drive reaction towards the products to compensate for the excess of PCl_3.

(b)   Removal of Cl_2 would cause the reaction to proceed toward the reactants to reestablish equilibrium and make up for the loss of Cl_2.

(c)   Removal of PCl_5 would cause the reaction to proceed toward the products to compensate for the loss of PCl_5.

(d)   A decrease in the volume of the container would cause the pressure to increase, and the reaction would favor the side which has the least number of moles of gas, namely the products.

(e)   Addition of He (an inert gas), without a change in the size of the container, would increase the pressure. There would be no effect on the position of equilibrium, however.

13.17 None of the above will effect the equilibrium constant for the reaction. The only change which will effect the equilibrium constant is a change in temperature.

13.18 (a) decrease H_2
     (b) increase H_2
     (c) no change. A catalyst does not affect the position of equilibrium.
     (d) decrease H_2
     (e) Decreasing the volume of the container will increase the pressure, and the reaction will favor the side with the least number of moles of gas. Since there is an equal number of moles of gas on both sides, no net change will be observed. The concentration of H_2 will remain the same.

13.19 None of the changes in Question 13.19 will affect the equilibrium constant except the change in temperature. Increasing the temperature will increase K.

13.20 (a) increase amount of NO   (c) increase amount of NO
     (b) decrease amount of NO   (d) no effect

13.21 (a) decrease amount of NH_3
     (b) increase amount of NH_3
     (c) decrease amount of NH_3
     (d) decrease amount of NH_3 (Note that H_2O is <u>liquid</u>, not gas.)

13.22

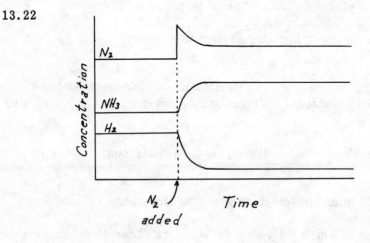

13.23 (a) no change   (c) decrease amount of $CaCO_3$
     (b) increase amount of $CaCO_3$ (d) increase amount of $CaCO_3$

CHAPTER 14

ACIDS AND BASES

Rationale

Acids and bases were introduced in Chapter 6 from essentially an Arrhenius point of view. The Arrhenius concept is reviewed here prior to an entire chapter devoted to aqueous acid-base equilibria.

One of the primary goals of this chapter is to show that acid-base phenomena are not restricted to water solutions alone. This is accomplished by expanding the Arrhenius definition to include any proton transfer (Brønsted-Lowry) reaction. The acid-base concept is then further extended to free acids and bases from the proton. Finally, the generalized solvent-system approach is examined to show that solvent involvement in acid-base reactions is not restricted to aqueous systems.

Objectives

At the conclusion of this chapter, students should be able to:

Recognize acids and bases by applying the Arrhenius, Brønsted-Lowry and Lewis definitions.

Define: acid anhydride, basic anhydride, amphoteric, conjugate acid-base pair, leveling effect, leveling solvent, differentiating solvent, hydrolysis, and solvolysis.

Illustrate the Brønsted-Lowry acid-base theory and identify conjugate acid-base pairs.

Distinguish between nucleophilic and electrophilic displacement.

Classify certain substances as acids or bases in solvents other than water by applying the solvent system approach.

Answers to Questions

14.1   Neutralization, reaction with indicators, and catalysis are three properties that are characteristic of acids and bases in general.

14.2   An acid is any substance that can increase the concentration of hydronium ion, H_3O^+, in aqueous solution.
A base is a substance that increases the hydroxide ion concentration in aqueous solution.

14.3   (a) $P_4O_{10} + 6H_2O \longrightarrow 4H_3PO_4$ (Acid)

                    $H_3PO_4 \rightleftharpoons H^+ + H_2PO_4^-$

      (b) $CaO + H_2O \longrightarrow Ca^{2+} + 2OH^-$ (Base)

      (c) $NH_3OH + H_2O \rightleftharpoons NH_4OH^+ + OH^-$ (Base)

      (d) $HBr + H_2O \longrightarrow Br^- + H_3O^+$ (Acid)

      (e) $H_2O + H_2O \rightleftharpoons OH^- + H_3O^+$ (Amphoteric)

      (f) $N_2O_5 + H_2O \longrightarrow 2HNO_3$ (Acid)

      (g) $Ba(OH)_2 \longrightarrow Ba^{++} + 2OH^-$ (Base)

14.4 $H_3O^+ + OH^- \longrightarrow 2H_2O$

14.5   <u>Acid anhydride</u> - nonmetal oxides that react with H_2O to yield acidic solutions.
       e.g. $CO_2 + H_2O \rightleftharpoons H_2CO_3$

                 $H_2CO_3 + H_2O \rightleftharpoons H_3O^+ + HCO_3^-$

<u>Basic anhydride</u> - metal oxides that react with H_2O to give corresponding hydroxides.
       e.g. $BaO + H_2O \longrightarrow Ba(OH)_2$

<u>Brønsted-Lowry definition of acids and bases</u>
Acid: substance that is able to donate a proton to some other substance.
Base: substance that is able to accept a proton from an acid.

The Brønsted-Lowry definition is less restrictive than the Arrhenius concept because it recognizes acid-base phenomena in other than just aqueous solutions.

14.6   Acid/base conjugate pairs (acid written first in each pair)

      (a) $HC_2H_3O_2$, $C_2H_3O_2^-$ and H_2O, OH^-

      (b) HF, F^- and NH_4^+, NH_3

(c) $Zn(OH)_2$, $ZnO_2{}^{2-}$ and H_2O, OH^-

(d) $Al(H_2O)_6{}^{3+}$, $Al(H_2O)_5OH^{2+}$ and H_2O, OH^-

(e) $N_2H_5{}^+$, N_2H_4 and H_2O, OH^-

(f) NH_3OH^+, NH_2OH and HCl, Cl^-

(g) OH^-, O^{2-} and H_2O, OH^-

(h) H_2, H^- and H_2O, OH^-

(i) NH_3, $NH_2{}^-$ and N_2H_4, $N_2H_3{}^-$

(j) HNO_3, $NO_3{}^-$ and $H_3SO_4{}^+$, H_2SO_4

14.7 Acid-base conjugate pairs (acids written first in each pair)

(a) $HClO_4$, $ClO_4{}^-$ and $N_2H_5{}^+$, N_2H_4

(b) H_3PO_3, $H_2PO_3{}^-$ and H_2SO_3, $HSO_3{}^-$

(c) $C_5H_5NH^+$, C_5H_5N and $(CH_3)_3NH^+$, $(CH_3)_3N$

(d) H_2O, OH^- and $HCO_3{}^-$, $CO_3{}^{2-}$

(e) $HCHO_2$, $CHO_2{}^-$ and $HC_7H_5O_2$, $C_7H_5O_2{}^-$

(f) $H_2C_2O_4$, $HC_2O_4{}^-$ and $CH_3NH_3{}^+$, CH_3NH_2

(g) H_2CO_3, $HCO_3{}^-$ and H_3O^+, H_2O

(h) C_2H_5OH, $C_2H_5O^-$ and NH_3, $NH_2{}^-$

(i) $N_2H_5{}^+$, N_2H_4 and HNO_2, $NO_2{}^-$

(j) H_2CN^+, HCN and H_2SO_4, $HSO_4{}^-$

14.8 $2H_2O \rightleftharpoons H_3O^+ + OH^-$

$2NH_3 \rightleftharpoons NH_4{}^+ + NH_2{}^-$

$2HCN \rightleftharpoons H_2CN^+ + CN^-$

14.9 acid, $(CH_3)_2NH_2{}^+$; base, $(CH_2)_2N^-$

14.10 (d) < (c) < (a) < (b)

14.11 (c) < (d) < (b) < (a)

14.12 $PH_2{}^-$ is a stronger base than HS^-.

14.13 CN^- is a stronger base than $NO_2{}^-$.

14.14 Ammonia

14.15 (b) < (a) < (d) < (c) < (e)

14.16 in the direction of $NH_4^+ + OCl^-$

14.17 $SO_4^{2-} < C_2H_3O_2^- < HCO_3^- < OCl^- < NH_3$

14.18 In comparing HCl and HBr in an aqueous medium, the difficulty that arises is that both appear to dissociate 100%; hence it is impossible to compare their strengths due to this leveling effect. Acetic acid would enable one to determine the relative strengths of these two acids.

14.19 Acetic acid would give smaller K; ammonia would give larger K.

14.20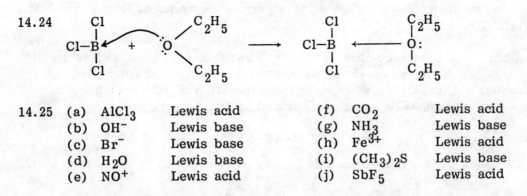

14.21 An acid that is stronger than H_2O will force its proton onto the H_2O molecule, forming H_3O^+ and the conjugate base of the other acid.

14.22 hydroxide ion

14.23 Lewis base: substance that can donate a pair of electrons to the formation of a covalent bond.
       Lewis acid: a substance that can accept a pair of electrons to form a covalent bond.

14.24

14.25 (a) $AlCl_3$ Lewis acid   (f) CO_2 Lewis acid
       (b) OH^- Lewis base   (g) NH_3 Lewis base
       (c) Br^- Lewis base   (h) Fe^{3+} Lewis acid
       (d) H_2O Lewis base   (i) $(CH_3)_2S$ Lewis base
       (e) NO^+ Lewis acid   (j) SbF_5 Lewis acid

14.26 NH_3, because on NF_3 the large electronegativity of fluorines makes the lone pair less available.

14.27 Because P is less electronegative than N, we would predict the $(CH_3)_3P:$ would be the stronger base.

14.28 (a) BCl_3 (b) Cr^{3+}

14.29 (a) Cl^-

(b)

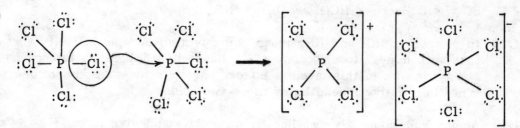

(c) electrophilic displacement

14.30

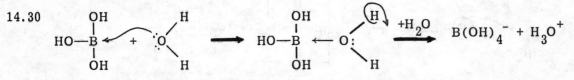

$B(OH)_3(H_2O)$ is acidic because the boron drains electron density away from the O—H bonds in the H_2O, thereby allowing the hydrogen to be more easily removed as H^+.

14.31

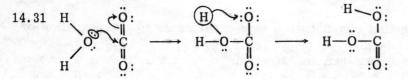

Water acts as a Lewis base by donating a pair of electrons to form a covalent bond with CO_2.

14.32 Nucleophile: A substance that in its reaction seeks a nucleus with which it can share a pair of electrons; i.e., a Lewis base.

Electrophile: A substance that seeks substances having electron pairs to which they can become bound; i.e., a Lewis acid.

14.33 $H_3N: \longrightarrow Ag^+ \longleftarrow :NH_3 \longrightarrow [H_3N-Ag-NH_3]^+$

NH_3 is a Lewis base which forms coordinate covalent bonds to the Lewis acid, Ag^+.

14.34 Acid: a substance that gives the cation that is also formed in the autoionization of the solvent.

Base: a substance that gives the anion that is also formed in the autoionization of the solvent.

14.35 (a) $HC_2H_3O_2 + H_2SO_4 \rightleftharpoons H_2C_2H_3O_2^+ + HSO_4^-$

(b) $HClO_4 + H_2SO_4 \rightleftharpoons ClO_4^- + H_3SO_4^+$

(c) $H_3SO_4^+ + HSO_4^- \rightleftharpoons 2H_2SO_4$

14.36 (a) $P_3N_5 + 7NH_3 \longrightarrow 3P(NH)(NH_2)_3$

(b) $Cl_2 + NH_3 \longrightarrow HCl + H_2NCl$

(c) $CH_3C(NH)(NHC_2H_5) + NH_3 \longrightarrow CH_3C(NH)(NH_2) + C_2H_5NH_2$

(d) $Zn + 2NH_4Cl \longrightarrow ZnCl_2 + H_2 + 2NH_3$

(e) $C(NH)(NH_2)_2 + 2NH_2^- \longrightarrow C(NH)_3^{2-} + 2NH_3$

(f) $SiCl_4 + 8NH_3 \longrightarrow Si(NH_2)_4 + 4NH_4Cl$

(g) $Zn + 2NH_3 + 2NH_2^- \longrightarrow Zn(NH_2)_4^{2-} + H_2$

(h) $Cu^{2+} + 4NH_3 \longrightarrow Cu(NH_3)_4^{2+}$

(i) $BCl_3 + 6NH_3 \longrightarrow 3NH_4Cl + B(NH_2)_3$

14.37 (a) Base, because it gives CN^-, the solvent anion.

(b) $H_2SO_4 + HCN \rightleftharpoons H_2CN^+ + HSO_4^-$

(c)

(d) $(CH_3)_3NH^+ + CN^- + H_2CN^+ + HSO_4^- \longrightarrow 2HCN + (CH_3)_3NH^+ + HSO_4^-$

(e) $H_2CN^+ + CN^- \longrightarrow 2HCN$

14.38 See Section 14.6.

14.39 (a) The highly charged Cr^{3+} ion polarizes the O–H bonds of the water molecules attached to it, thereby making it easier for them to be transferred to neighboring H_2O molecules.

(b) $Cr(H_2O)_6^{3+} + H_2O \rightleftharpoons Cr(H_2O)_5OH^{2+} + H_3O^+$

(c) Cr^{3+} is better able to polarize the H_2O molecules than is Cr^{2+}.

CHAPTER 15

ACID-BASE EQUILIBRIA IN AQUEOUS SOLUTION

Rationale

This chapter provides a thorough discussion of ionic equilibria in acid-base reactions. Students generally have a difficult time with this material and sufficient flexibility should be built into your lecture schedule so that you can slow down a little when students near their limit of saturation. Encourage students to set up the data table to organize their thinking. Remind them once again that only quantities from the equilibrium column can be inserted into the K_{eq} expression.

Objectives

At the conclusion of this chapter, students should be able to:

Define: ionization constant, pH, pOH, pK, weak acid or base, polyprotic acid, buffer, hydrolysis, and equivalence point.

Calculate the pH of a solution of a strong acid or base.

Perform calculations involving dissociation of weak acids and bases.

Perform calculations involving polyprotic acids.

Explain, qualitatively, how a buffer functions.

Calculate the pH of a buffer and the effects of additions of small amounts of strong acids or bases.

Determine amounts of reagents required to prepare buffers of specified pH.

Explain, qualitatively, why solutions of some salts undergo hydrolysis.

Perform hydrolysis calculations.

Calculate the pH at various points during a titration of (a) strong acid – strong base, (b) strong acid – weak base, (c) weak acid – strong base.

Choose an appropriate indicator for an acid–base titration.

Answers to Questions

15.1   In pure water the $[H^+] = 1 \times 10^{-7}$. In the presence of an acid the dissociation of water is suppressed (Le Châtelier's principle) and the $[H^+]$ contributed from the H_2O is less than 10^{-7} M. Only in <u>very</u> dilute acid solutions must the dissociation of water be taken into account.

15.2   pH = -log $[H^+]$;   pOH = -log $[OH^-]$; see Page 466.

15.3   (a) acidic (b) basic (c) neutral (d) acidic (e) basic

15.4   pOH = 2.25, pH = 8.25, pOH = 7.00, pOH = 10.43, pH = 3.54
       (e) < (b) < (c) < (d) < (a)

15.5   (a) $HC_7H_5O_2 \rightleftharpoons H^+ + C_7H_5O_2^-$ $K_{eq} = \dfrac{[H^+][C_7H_5O_2^-]}{[HC_7H_5O_2]}$

       (b) $N_2H_5 + H_2O \rightleftharpoons N_2H_5^+ + OH^-$ $K_{eq} = \dfrac{[N_2H_5^+][OH^-]}{[N_2H_4]}$

       (c) $HCHO_2 \rightleftharpoons H^+ + CH_2O^-$ $K_{eq} = \dfrac{[H^+][CH_2O^-]}{[HCHO_2]}$

       (d) $HC_8H_{11}N_2O_3 \rightleftharpoons H^+ + C_8H_{11}N_2O_3^-$ $K_{eq} = \dfrac{[H^+][C_8H_{11}N_2O_3^-]}{[HC_8H_{11}N_2O_3]}$

       (e) $C_6H_5N + H_2O \rightleftharpoons C_6H_5NH^+ + OH^-$ $K_{eq} = \dfrac{[C_6H_5NH^+][OH^-]}{[C_6H_5N]}$

15.6 $H_2C_6H_6O_6 \rightleftharpoons H^+ + HC_6H_6O_6^-$ $K_{a_1} = \dfrac{[H^+][HC_6H_6O_6^-]}{[H_2C_6H_6O_6]}$

$$HC_6H_6O_6^- \rightleftharpoons H^+ + C_6H_6O_6^{2-} \qquad K_{a_2} = \frac{[H^+][C_6H_6O_6^{2-}]}{[HC_6H_6O_6^-]}$$

15.7 $\quad H_3C_6H_5O_7 \rightleftharpoons H_2C_6H_5O_7^- + H^+ \qquad K_{a_1} = \dfrac{[H_2C_6H_3O_7^-][H^+]}{[H_3C_6H_5O_7]}$

$\qquad H_2C_6H_5O_7^- \rightleftharpoons HC_6H_5O_7^{2-} + H^+ \qquad K_{a_2} = \dfrac{[HC_6H_5O_7^{2-}][H^+]}{[H_2C_6H_5O_7^-]}$

$\qquad HC_6H_5O_7^{2-} \rightleftharpoons C_6H_5O_7^{3-} + H^+ \qquad K_{a_3} = \dfrac{[C_6H_5O_7^{3-}][H^+]}{[HC_6H_5O_7^{2-}]}$

15.8 A buffer is any solution that contains both a weak acid and a weak base and has the property that the addition of small quantities of a strong acid are neutralized by the weak base while small quantities of a strong base are neutralized by the weak acid.

(a) $NaCHO_2$ provides weak conjugate base CHO_2^-; $HCHO_2$ is the weak acid.

(b) C_5H_5N provides the weak base; C_5H_5NCl provides weak acid $C_5H_5NH^+$.

(c) $NH_4C_2H_3O_2$ provides weak base $C_2H_3O_2^-$ and the weak acid NH_4^+.

(d) $NaHCO_3$ provides weak base HCO_3^-; HCO_3^- is <u>both</u> a weak acid and a weak base.

15.9 No. HCl is a strong acid and is completely dissociated. Cl^- is a very weak conjugate base and cannot neutralize another "acid."

15.10 For additions of strong acid, $\quad HPO_4^{2-} + H_3O^+ \longrightarrow H_2PO_4^- + H_2O$

For additions of strong base, $\quad H_2PO_4^- + OH^- \longrightarrow HPO_4^{2-} + H_2O$

15.11 Hydrolysis: When a salt dissolves in water, it dissociates fully to produce cations and anions that may subsequently react chemically with the solvent in a process called hydrolysis; e.g., the cation of the salt undergoes the equation $M^+ + H_2O \rightleftharpoons MOH + H^+$ while an anion reacts according to the equation $X^- + H_2O \rightleftharpoons HX + OH^-$.

(a) neutral (b) acidic (c) basic (d) acidic

15.12 $NaC_4H_7O_2$ most basic

        $C_6H_5NH_3NO_3$ most acidic

15.13 The second hydrolysis reaction occurs to a neglibible extent compared to the first. e.g.

$$K_{h_1} = \frac{K_w}{K_{a_2}} = \frac{1 \times 10^{-14}}{1 \times 10^{-7}} = 1 \times 10^{-7}$$

$$K_{h_2} = \frac{K_w}{K_{a_1}} = \frac{1 \times 10^{-14}}{1.5 \times 10^{-2}} = 7 \times 10^{-13}$$

15.14 Yes - due to hydrolysis.

15.15 The endpoint occurs when the indicator changes color, which may or may not occur at the time that the equivalence point is reached.

15.16 Acid and basic forms of an indicator differ in color ($HIn \rightleftharpoons H^+ + In^-$). Therefore, depending on the pH range of your indicator, it will change to HIn in acid solution and In^- in basic solution. This color change ideally should correspond to the endpoint of the titration. If too much indicator is added, it may interfere with the endpoint because it will react with the base in the titration.

15.17 Thymol Blue or Phenolphthalein
Congo Red pH range is too low. It would change color before the equivalence point is reached.

15.18 No. pH range for the color change is too low.

CHAPTER 16

SOLUBILITY AND COMPLEX ION EQUILIBRIA

Rationale

This chapter concludes the discussion of ionic equilibrium with a treatment of solubility product and complex ions. Once again, encourage students to organize their problem solving by constructing the data table. This is particularly important with problems involving the common-ion effect similar to Example 16.9 on Page 509.

Objectives

Upon completing this chapter, students should be able to:

Compute the solubility product constant of insoluble salts given their solubilities, and vice versa.

Determine whether a precipitate will form in a solution.

Predict, mathematically, which ion will precipitate when a precipitating agent is added to a solution of two or more ions.

Apply the common ion effect to determine the solubility of salts in a solution that has an ion in common with the salt.

Determine the molar solubility of salts in solutions that form complex ions with the ions of the "insoluble" salt.

Answers to Questions

16.1   The concentration of a solid in the solid is a constant. Thus, the concentration of the solid can be included in the equilibrium constant.

16.2   (a) $Ag_2S(s) \rightleftharpoons 2Ag^+ + S^{2-}$ $K_{sp} = [Ag^+]^2[S^{2-}]$

      (b) $CaF_2(s) \rightleftharpoons Ca^{2+} + 2F^-$ $K_{sp} = [Ca^{2+}][F^-]^2$

      (c) $Fe(OH)_3(s) \rightleftharpoons Fe^{3+} + 3OH^-$ $K_{sp} = [Fe^{3+}][OH^-]^3$

      (d) $MgC_2O_4(s) \rightleftharpoons Mg^{2+} + C_2O_4^{2-}$ $K_{sp} = [Mg^{2+}][C_2O_4^{2-}]$

      (e) $Bi_2S_3(s) \rightleftharpoons 2Bi^{3+} + 3S^{2-}$ $K_{sp} = [Bi^{3+}]^2[S^{2-}]^3$

      (f) $BaCO_3(s) \rightleftharpoons Ba^{2+} + CO_3^{2-}$ $K_{sp} = [Ba^{2+}][CO_3^{2-}]$

16.3   (a) $K_{sp} = [Pb^{2+}][F^-]^2$ (d) $K_{sp} = [Li^+]^2[CO_3^{2-}]$

      (b) $K_{sp} = [Cu^+]^2[S^{2-}]$ (e) $K_{sp} = [Ca^{2+}][IO_3^-]^2$

      (c) $K_{sp} = [Fe^{2+}]^3[PO_4^{3-}]^2$ (f) $K_{sp} = [Ag^+]^2[Cr_2O_7^{2-}]$

16.4   The solubility product must be exceeded by the ion product.

16.5   Complex ion is an ion containing a metal ion combined with one or more molecules or ions to produce a more complex species; e.g., $CuCl_4^{2-}$.
Ligands are substances that combine with metal ions and are usually molecules and ions that are Lewis bases; e.g., H_2O, Cl^-, CN^-.

16.6   Consider the three equilibria:

    (1) $NH_4^+ + H_2O \rightleftharpoons NH_3 + H_3O^+$

    (2) $Mg(OH)_2(s) \rightleftharpoons Mg^{2+} + 2OH^-$

    (3) $H_3O^+ + OH^- \rightleftharpoons 2H_2O$

When NH_4^+ is added to the solution, the position of equilibrium in Equation (1) shifts to the right, producing H_3O^+. This H_3O^+ reacts with OH^- according to Equation (3). When OH^- is removed from the solution, $Mg(OH)_2$ dissolves in an attempt to replace it.

16.7 $AgI_2^- \rightleftharpoons AgI + I^-$

       $AgI \rightleftharpoons Ag^+ + I^-$

       _____

       $AgI_2^- \rightleftharpoons Ag^+ + 2I^-$

$$K_{eq} = \frac{[Ag^+][I^-]^2}{[AgI_2^-]} \qquad \text{This } K_{eq} \text{ has units:} \qquad \frac{\left(\dfrac{mole}{liter}\right)\left(\dfrac{mole}{liter}\right)^2}{(mole/liter)}$$

$$K_{eq} = \frac{(moles\ Ag^+)(moles\ I^-)^2}{(moles\ AgI_2^-)}\ \frac{1}{liter^2}$$

From the preceding equations we see that in order to maintain a constant K_{eq}, as the volume (liter) increases, the moles of Ag^+ and I^- must increase; i.e., AgI_2^- decomposes.

When the ion product, $[Ag^+][I^-]$ exceeds the K_{sp} of AgI, a precipitate of AgI will form.

16.8   As applied to solubility equilibria, a salt is less soluble in water if the solution contains one of the ions of the salt (the common ion). In general, a common ion suppresses the dissociation of a weak electrolyte.

16.9   (a)
$$K_{form} = \frac{[AgCl_2^-]}{[Ag^+][Cl^-]^2}$$

(b)
$$K_{form} = \frac{[Ag(S_2O_3)_2^{3-}]}{[Ag^+][S_2O_3^{2-}]^2}$$

(c)
$$K_{form} = \frac{[Zn(NH_3)_4^{2+}]}{[Zn^{2+}][NH_3]^4}$$

16.10   (a)
$$K_{inst} = \frac{[Fe^{2+}][CN^-]^6}{[Fe(CN)_6^{4-}]}$$

(b)
$$K_{inst} = \frac{[Cu^{2+}][Cl^-]^4}{[CuCl_4^{2-}]}$$

(c)
$$K_{inst} = \frac{[Ni^{2+}][NH_3]^6}{[Ni(NH_3)_6^{2+}]}$$

CHAPTER 17

ELECTROCHEMISTRY

Rationale

The chapter begins with electrolysis and applications of Faraday's laws. The discussion provides an opportunity to describe several important commercial applications of electrolysis. Next, galvanic cells are introduced and the concepts of reduction potentials are developed. The uses of a table of reduction potentials are discussed on Pages 536 to 540.

The relationship of the cell potential to ΔG is examined next. The discussion relates back to Chapter 11 where ΔG was associated with the maximum available useful work that can be obtained from a system. It is here that viewing the EMF as the energy available per coulomb permits the development of the important equation, $\Delta G = -n\mathscr{F}\mathscr{E}$.

In examining the effect of concentration on cell potential we again harken back to a previous chapter (Section 13.3) which permits the simple derivation of the Nernst equation.

The last three sections provide some practical applications of galvanic cells. Sections 17.11 and 17.12 deal with the electrical measurement of concentrations; Section 17.13 with familiar (and perhaps some unfamiliar) batteries.

Objectives

At the conclusion of this chapter, students should be able to:

Distinguish between electrolytic and galvanic cells, metallic and electrolytic conduction, cathode and anode in electrolytic and galvanic cells.

Define: electrolysis, cell reaction, electroplating, faraday, coulometer, electromotive force, and cell potential.

Describe the electrode reactions that occur during the electrolysis of: molten sodium chloride, aqueous sodium chloride, aqueous copper sulfate, aqueous copper chloride, aqueous sodium sulfate, aluminum ore, molten magnesium chloride, and the refining of copper.

Calculate the amount of chemical change that will occur during electrolysis, in single solutions and in coulometers.

Define: standard reduction potentials, reference electrode, thermodynamic equilibrium constants, concentration cells, ion selective electrode and fuel cells.

Predict the cell potential with knowledge of the reduction potentials of the two half-reactions.

Use reduction potentials to predict the spontaneity of redox reactions.

Determine thermodynamic equilibrium constants, solubility product constants and pH by application of the Nernst equation.

Predict the effect of concentration changes on the potential of a cell.

Calculate the voltage of certain concentration cells.

Write chemical equations for the cell reactions in the lead storage battery, the Leclanché dry cell, the alkaline battery, the silver oxide battery, the nickel-cadmium battery, and the hydrogen-oxygen fuel cell.

Answers to Questions

17.1   (a) Electrolytic cells convert electrical energy into chemical energy and galvanic cells convert chemical energy into electrical energy.

        (b) In metallic conduction electrons move through a metal. In electrolytic conduction, ions move through a solution.

        (c) oxidation - loss of electrons (occurs at anode)
              reduction - gain of electrons (occurs at cathode)

17.2   Oxidation - reduction maintains electrical neutrality.

17.3   Reduction takes place at the cathode and oxidation takes place at the anode.

17.4 $2H_2O \longrightarrow O_2 + 4H^+ + 4e^-$ (oxidation)

       $2H_2O + 2e^- \longrightarrow H_2 + 2OH^-$ (reduction)

17.5 The products would be O_2 at the anode and H_2 at the cathode due to the oxidation and reduction of water.

17.6 Na_2SO_4 and H_2SO_4 are needed to maintain electrical neutrality. H^+ and OH^- ions are produced during the oxidation and reduction of water. Ions of opposite charge are required in the vicinity of these ions to "neutralize" their charge.

17.7 (a) cathode: $2H_2O + 2e^- \longrightarrow H_2 + 2OH^-$

anode: $2Cl^- \longrightarrow Cl_2 + 2e^-$

net: $2H_2O + 2Cl^- \longrightarrow H_2 + Cl_2 + 2OH^-$

(b) In a stirred solution, Cl_2 reacts with OH^-

$$Cl_2 + 2OH^- \longrightarrow Cl^- + OCl^- + H_2O$$

Net reaction is: $Cl^- + H_2O \longrightarrow OCl^- + H_2$

17.8 Advantage: NaOH not contaminated by Cl^- is produced.
Disadvantage: possibility for mercury pollution.

17.9 It keeps Na and Cl_2 apart so they cannot react with each other to reform NaCl.

17.10 Cryolite reduces the melting temperature of Al_2O_3 from 2000°C to 1000°C.

17.11 H_2O is more easily reduced than Al^{3+}.

17.12 $Mg^{2+}(aq) + CaO(s) + H_2O \longrightarrow Mg(OH)_2(s) + Ca^{2+}(aq)$

$Mg(OH)_2(s) + 2HCl \longrightarrow MgCl_2(s) + 2H_2O$

$MgCl_2(\ell) \xrightarrow{\text{electrical}\atop\text{energy}} Mg(\ell) + Cl_2(g)$

17.13 For the electrolytic purification of copper see Page 526. The process is economically feasible because the impurities, silver, gold and platinum, can be sold for enough money to pay for the electricity required for the electrolysis.

17.14 A faraday is the amount of electricity that must be supplied to a cell in order to supply one mole of electrons.

17.15 A coulometer is an apparatus that allows us to experimentally determine the weight of a substance that has been deposited on an electrode during electrolysis. Coulometers can be very accurate and can be hooked up in series, which allows for the passage of the same number of faradays through two or more such cells. See Figure 17.10.

17.16 In galvanic cells the anode is negative and the cathode is positive. In electrolytic cells the anode is positive and the cathode is negative.

17.17 The electron flow takes place on the surface of the zinc. Electrons are removed from the zinc and are picked up by the Cu^{2+} ions that are in the vicinity. The energy change appears as heat.

17.18 The salt-bridge is needed in order to maintain electrical neutrality.

17.19

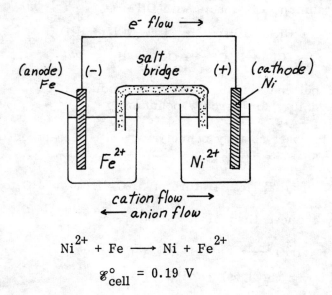

$$Ni^{2+} + Fe \longrightarrow Ni + Fe^{2+}$$

$$\mathscr{E}^{\circ}_{cell} = 0.19 \text{ V}$$

17.20 One volt is equal to one joule per coulomb; it is a measure of the energy that is capable of being extracted from the flowing electric charge. One ampere is equivalent to one coulomb per second, and is a measure of the current.

17.21 (a) $2Al + 3Ni^{2+} \longrightarrow 3Ni + 2Al^{3+}$

(b) $3PbO_2 + 2Cr^{3+} + 3SO_4^{2-} + H_2O \longrightarrow 3PbSO_4 + 2H^+ + Cr_2O_7^{2-}$

(c) $2Ag^+ + Pb \longrightarrow 2Ag + Pb^{2+}$

(d) $Cl_2 + Mn^{2+} + 2H_2O \longrightarrow 2Cl^- + 4H^+ + MnO_2$

(e) $2H^+ + Mn \longrightarrow Mn^{2+} + H_2$

17.22 Spontaneous reactions are (a), (d), (e). The other reactions are spontaneous in the reverse direction.

17.23 Spontaneous reactions are (c) and (d). The others are spontaneous in the reverse direction.

17.24 In a galvanic cell, the negative electrode is the anode; the positive electrode is the cathode. These can be determined with a voltmeter. Another method would be to do a chemical analysis of the products formed at each electrode. Oxidation occurs at the anode, reduction at the cathode.

17.25 (a) Ca^{2+} (b) F_2 (c) H_2O (d) $S_2O_8^{2-}$ (e) Br_2

17.26 (a) ClO_3^- (b) $Cr_2O_7^{2-}$ (c) MnO_4^- (d) PbO_2

17.27 (a) Fe (b) Mg (c) I^- (d) SO_4^{2-} (e) Mn

17.28 (a) Na (b) Cl_2 (c) Cu (d) Sn (e) H_2

17.29 A concentration cell is one in which the cathode and anode compartments contain the same electrode materials, but different concentrations of the ions.

17.30 (anode) $\quad Zn \longrightarrow Zn^{2+} + 2e^-$

(cathode) $\quad 2MnO_2 + 2NH_4^+ + 2e^- \longrightarrow Mn_2O_3 + 2NH_3 + H_2O$

17.31 The products formed at the electrodes diffuse away and the cell is rejuvenated.

17.32 anode: $\quad Zn + 2OH^- \longrightarrow Zn(OH)_2 + 2e^-$

cathode: $\quad 2MnO_2 + 2H_2O + 2e^- \longrightarrow 2MnO(OH) + 2OH^-$

net: $\quad Zn + 2MnO_2 + 2H_2O \longrightarrow Zn(OH)_2 + 2MnO(OH)$

Electrolyte is aqueous KOH.

17.33 anode: $\quad Zn + 2OH^- \longrightarrow Zn(OH)_2 + 2e^-$

cathode: $\quad Ag_2O + H_2O + 2e^- \longrightarrow 2Ag + 2OH^-$

net: $\quad Zn + Ag_2O + H_2O \longrightarrow Zn(OH)_2 + 2Ag$

17.34 (anode) $\quad Pb(s) + SO_4^{2-} \longrightarrow PbSO_4(s) + 2e^-$

(cathode) $\quad PbO_2(s) + 4H^+ + SO_4^{2-} + 2e^- \longrightarrow PbSO_4(s) + 2H_2O$

17.35 (anode) $Cd(s) + 2OH^- \longrightarrow Cd(OH)_2(s) + 2e^-$

 (cathode) $NiO_2(s) + 2H_2O + 2e^- \longrightarrow Ni(OH)_2(s) + 2OH^-$

17.36 Fuel cells produce electrical energy from gaseous fuels undergoing reaction in a carefully designed environment. The reactants in a fuel cell may be continually supplied so that energy can be withdrawn as long as the outside fuel supply is maintained.

17.37 Fuel cells operate under more nearly reversible conditions. Therefore, their thermodynamic efficiency is higher; i.e., more of the available energy can be used to do work.

CHAPTER 18

CHEMICAL PROPERTIES OF THE REPRESENTATIVE METALS

Rationale

This is the first of four chapters dealing with descriptive inorganic chemistry, and as its name suggests it is devoted to a discussion of the chemistry of the A-group metals. We begin by examining, in a general way, how metals are extracted from their compounds and what factors influence the choice of method. This discussion provides an opportunity to apply some of the principles learned in earlier chapters. Then we proceed with a discussion of specific properties of the A-group metals. Our goal here has not been to be encyclopedic, but rather to present the most important chemistry of these elements in an interesting fashion. In teaching this material, we suggest that you provide as many examples as possible that illustrate how the chemistry relates to common household substances and to products with which students may be familiar.

Objectives

When students have completed this chapter, they should be able to:

Describe how metals may be extracted from their ores.

Determine the temperature at which thermal decomposition of a compound becomes thermodynamically favorable.

Identify each of the metals in the A-groups of the periodic table.

Describe where the alkali metals are found and how they are obtained from their compounds.

Describe the physical properties of the alkali metals.

Describe the reactions of the alkali metals with: water, liquid ammonia, oxygen, and nitrogen.

Write equations for the reactions of the oxides of the alkali metals with water, and for the preparation of sodium bicarbonate by the Solvay process.

Describe the occurrence and preparation of the alkaline earth metals, especially the extraction of magnesium from sea water.

Compare the physical and chemical properties of the alkali and alkaline earth metals.

Discuss the properties and reactions of the oxides, hydroxides, carbonates, and sulfates of the alkaline earth metals.

Compare the reactions of the alkaline earth metals with water, oxygen, and nitrogen.

Describe the structure of beryllium chloride.

Describe the occurrence and preparation of the metals of Group IIIA.

Describe the structure of aluminum chloride and the amphoteric behavior of aluminum.

Define what an alum is.

Compare the stabilities of the higher and lower oxidation states of tin, lead, and bismuth.

Answers to Questions

18.1 They are usually found in the combined state. Sources are the oceans and land-based deposits of carbonates, sulfates, oxides, and sulfides.

18.2 It must have a small ΔH_f°.

18.3 In compounds, metals exist in positive oxidation states.

18.4 It has a large, negative ΔH_f°.

18.5 It is plentiful and inexpensive.

18.6 $2PbO + C \longrightarrow 2Pb + CO_2$

$PbO + H_2 \longrightarrow Pb + H_2O$

18.7 The chemical reducing agent itself would be even more difficult to prepare.

18.8 They tend to have relatively low melting points.

18.9

$$2NaCl(\ell) \xrightarrow{\text{electrolysis}} 2Na(\ell) + Cl_2(g)$$

$$2Al_2O_3(\ell) \xrightarrow[\text{cryolite}]{\text{electrolysis}} 4Al(\ell) + 3O_2(g)$$

18.10 Their oxides are basic. Oxidation states are zero and 1+.

18.11 Na and K. Francium is least abundant because it is radioactive with a short half-life.

18.12 They are relatively rare elements and compounds of Na and K usually serve just as well.

18.13 In the ocean; in salt deposits.

18.14 $MCl(\ell) + Na(g) \longrightarrow NaCl(\ell) + M(g)$ [M = K, Rb, Cs]

18.15 Cooling nuclear reactors; sodium vapor lamps.

18.16 (a) yellow   (b) red   (c) violet

18.17 View the flame through blue "cobalt glass."

18.18 $2Rb + 2H_2O \longrightarrow 2RbOH + H_2(g)$

18.19 Because of the very large hydration energy of the tiny Li^+ ion.

18.20 It dissolves, giving a blue solution. These solutions contain solvated electrons.

18.21 (a) $2Li + Br_2 \longrightarrow 2LiBr$

   $2Na + Br_2 \longrightarrow 2NaBr$

   (b) $2Li + S \longrightarrow Li_2S$

   $2Na + S \longrightarrow Na_2S$

   (c) $6Li + N_2 \longrightarrow 2Li_3N$

   $Na + N_2 \longrightarrow$ no reaction

18.22 $4Li + O_2 \longrightarrow 2Li_2O$

$2Na + O_2 \longrightarrow Na_2O_2$

$M + O_2 \longrightarrow MO_2$ [M = K, Rb, Cs]

18.23 KO_2

$4KO_2 + 2CO_2 \longrightarrow 2K_2CO_3 + 3O_2$

$2KO_2 + 2H_2O \longrightarrow 2KOH + O_2 + H_2O_2$

18.24   Because of hydrolysis

$O_2^{2-} + H_2O \longrightarrow HO_2^- + OH^-$

18.25   A metal that reacts with traces of O_2 and removes them from an otherwise inert atmosphere.

18.26   They are less expensive to produce.

18.27   Caustic soda, lye. Uses include: making soap; neutralizing acids; in drain cleaners; making detergents, pulp, and paper; removing sulfur from petroleum.

18.28 $Na_2CO_3 \cdot NaHCO_3 \cdot 2H_2O$

Solvay process:

$$CaCO_3 \xrightarrow{\text{heat}} CaO + CO_2$$

$$CO_2 + H_2O \longrightarrow H_2CO_3$$

$$H_2CO_3 + NH_3 \longrightarrow NH_4^+ + HCO_3^-$$

$$HCO_3^- + Na^+ \longrightarrow NaHCO_3$$

$$2NaHCO_3 \xrightarrow{\text{heat}} Na_2CO_3 + H_2O$$

$$NH_4Cl + CaO \longrightarrow CaCl_2 + NH_3 + H_2O$$

Net reaction: $2NaCl + CaCO_3 \longrightarrow Na_2CO_3 + CaCl_2$

18.29   Buffer: The HCO_3^- can neutralize both acids and bases.

$$HCO_3^- + H_3O^+ \longrightarrow H_2CO_3 + H_2O$$

$$HCO_3^- + OH^- \longrightarrow CO_3^{2-} + H_2O$$

Fire extinguisher:

$$2NaHCO_3(s) \xrightarrow{\text{heat}} Na_2CO_3(s) + H_2O(g) + CO_2(g)$$

18.30 K_2CO_3

$KOH + CO_2 \longrightarrow KHCO_3$

$2KHCO_3 \xrightarrow{heat} K_2CO_3 + H_2O + CO_2$

18.31 Li_2CO_3

18.32 Their oxides are basic and their ores are found in the earth. They are all larger.

18.33 Their hydration energies are larger than those of the alkali metals, which off-sets their larger ionization energies.

18.34 Calcium and magnesium in mineral deposits ($CaCO_3$, $CaSO_4 \cdot 2H_2O$, $CaCO_3 \cdot MgCO_3$, $MgCl_2 \cdot KCl \cdot H_2O$) and in the sea.

Radium is found in pitchblend - a uranium ore.

18.35 From dolomite:

$CaCO_3 \cdot MgCO_3 \xrightarrow{heat} CaO \cdot MgO + 2CO_2$

$CaO + H_2O \longrightarrow Ca^{2+} + 2OH^-$

$MgO + H_2O \longrightarrow Mg(OH)_2$

$Mg(OH)_2 + 2HCl \longrightarrow MgCl_2 + 2H_2O$

$MgCl_2(\ell) \xrightarrow{electrolysis} Mg(\ell) + Cl_2(g)$

From sea water:

$CaO + H_2O + Mg^{2+} \longrightarrow Ca^{2+} + Mg(OH)_2$

$Mg(OH)_2 + 2HCl \longrightarrow MgCl_2 + 2H_2O$

$MgCl_2(\ell) \xrightarrow{electrolysis} Mg(\ell) + Cl_2(g)$

18.36 Calcining: heating a substance strongly.
Lime is CaO.

$CaO + H_2O \longrightarrow Ca(OH)_2$

It is an inexpensive strong base.

18.37 $2Mg + O_2 \longrightarrow 2MgO + light$

18.38 (a) brick red (b) crimson (c) yellowish-green

18.39 Their oxides provide a water-insoluble protective film that prevents further oxidation. Ca, Sr, and Ba react with water.

18.40 $2Mg + O_2 \longrightarrow 2MgO$

$Mg + S \longrightarrow MgS$

$3Mg + N_2 \longrightarrow Mg_3N_2$

18.41 $Be + 2H^+ \longrightarrow Be^{2+} + H_2$

$Be + 2H_2O + 2OH^- \longrightarrow Be(OH)_4^{2-} + H_2(g)$

18.42 See Figure 18.7.
Formation of the additional two Be—Cl coordinate covalent bonds suggests that the Be in $BeCl_2$ seeks additional electrons to complete its octet.

18.43 Because Be is so very small and highly charged, electron density is drawn toward Be^{2+}, making bonds covalent.
Organomagnesium compounds contain portions of organic molecules covalently bonded to Mg.

18.44 Solubility of the hydroxides increases and the sulfates decrease from top to bottom in Group IIA.

18.45 To make lime, as a mild abrasive, an antacid, chalk.

18.46 $Mg(OH)_2$ suspended in H_2O

18.47 $CaSO_4 \cdot 2H_2O$

$CaSO_4 \cdot 2H_2O \xrightarrow{heat} CaSO_4 \cdot 1/2H_2O + 3/2H_2O$

$CaSO_4 \cdot 1/2H_2O + 3/2H_2O \longrightarrow CaSO_4 \cdot 2H_2O$

18.48 It is opaque to X rays.

18.49 $MgSO_4 \cdot 7H_2O$

18.50 Many are amphoteric and form covalent compounds.

18.51 IIIA: 1+, 3+
IVA: 2+, 4+
VA: 3+, 5+

18.52 Ga, In, Tl, Sn, Pb, Bi
Lower oxidation states become more stable going down a group because the

energy needed to remove additional electrons can't be recovered by forming additional bonds. This is because bond strength decreases as the atoms become larger.

18.53 Bauxite, Al_2O_3

$$Al_2O_3(s) + \text{impurities (s)} \xrightarrow{OH^-} AlO_2^-(aq) + \text{impurities (s)}$$
$$AlO_2^- + H_3O^+ \longrightarrow Al(OH)_3(s)$$
$$2Al(OH)_3 \xrightarrow{heat} Al_2O_3 + 3H_2O$$

18.54 Structural metal, kitchen utensils, automobiles, aircraft, beverage cans, aluminum foil.

18.55 Tin: $\quad SnO_2 + C \longrightarrow Sn + CO_2$

Lead: $\quad 2PbS + 3O_2 \longrightarrow 2PbO + 2SO_2$
$$2PbO + C \longrightarrow 2Pb + CO_2$$

Bismuth: $\quad 2Bi_2S_3 + 9O_2 \longrightarrow 2Bi_2O_3 + 6SO_2$
$$2Bi_2O_3 + 3C \longrightarrow 4Bi + 3CO_2$$

18.56 Iron is more easily oxidized than tin. A galvanic cell is established in which iron is the anode.

18.57 Allotropes are different forms of the same element. Tin exhibits allotropism.

18.58 It expands slightly when its liquid freezes. Woods metal is an alloy of 50% Bi, 25% Pb, 12.5% Sn, and 12.5% Cd. It has a low melting point (70°C) and is used in fuses and in triggering mechanisms for automatic sprinkler systems.

18.59 The Al_2O_3 on its surface protects the metal beneath. Forming an amalgam with the Al surface prevents Al_2O_3 from adhering and causes rapid corrosion.

18.60 $2Al + 6H^+ \longrightarrow 2Al^{3+} + 3H_2(g)$

$2Al + 2OH^- + 2H_2O \longrightarrow 2AlO_2^- + 3H_2(g)$

18.61 γ-Al_2O_3 is quite reactive; α-Al_2O_3 is quite inert. Gems composed of Al_2O_3 are ruby, sapphire, topaz, amethyst.

18.62 $Fe_2O_3(s) + 2Al(s) \longrightarrow Al_2O_3(s) + 2Fe(\ell) + \text{heat}$

18.63 See figure on Page 581.

18.64 Hydrolysis: $Al(H_2O)_6^{3+} + H_2O \rightleftharpoons Al(H_2O)_5OH^{2+} + H_3O^+$

18.65 $Al(H_2O)_6^{3+} + OH^- \longrightarrow Al(H_2O)_5OH^{2+} + H_2O$

$Al(H_2O)_5OH^{2+} + OH^- \longrightarrow Al(H_2O)_4(OH)_2^+ + H_2O$

$Al(H_2O)_4(OH)_2^+ + OH^- \longrightarrow Al(H_2O)_3(OH)_3(s) + H_2O$

$Al(H_2O)_3(OH)_3(s) + OH^- \longrightarrow Al(H_2O)_2(OH)_4^- + H_2O$

18.66 AlO_2^- and $Al(H_2O)_2(OH)_4^-$

18.67 When solutions of $Al_2(SO_4)_3$ are made basic, hydrated aluminum hydroxide precipitates. As it settles, it carries fine sediment and bacteria with it.

18.68 A double salt of general formula $M^+M^{3+}(SO_4)_2 \cdot 12H_2O$. Example is $NaAl(SO_4)_2 \cdot 12H_2O$. This is the alum in baking powders. It evolves CO_2 from $NaHCO_3$ because the aluminum ion hydrolyzes, releasing H_3O^+ which reacts with the $NaHCO_3$.

18.69 $Sn + HNO_3 \longrightarrow SnO_2 + 4NO_2 + 2H_2O$

$3Pb + 8HNO_3 \longrightarrow 3Pb(NO_3)_2 + 2NO + 4H_2O$

$Sn + 2Cl_2 \longrightarrow SnCl_4$

$Pb + Cl_2 \longrightarrow PbCl_2$

In each case, tin gives the 4+ state; lead gives the 2+ state.

18.70 Bi^{5+} is such a powerful oxidizing agent it would oxidize Cl^- to Cl_2.

18.71 $Sn + 2OH^- + 2H_2O \longrightarrow Sn(OH)_4^{2-} + H_2(g)$

$Pb + 2OH^- + 2H_2O \longrightarrow Pb(OH)_4^- + H_2(g)$

18.72 PbO (litharge); used in pottery glazes and making lead crystal.
Pb_3O_4 (red lead); used in corrosion inhibiting paint.
PbO_2; cathode material in the lead storage battery.

18.73 Reaction of Pb^{2+} with airborn H_2S gives black PbS.

18.74 Cosmetics and pharmaceuticals. Bismuthyl ion is BiO^+.

CHAPTER 19

THE CHEMISTRY OF SELECTED NONMETALS, PART I:
HYDROGEN, CARBON, OXYGEN, AND NITROGEN

Rationale

In introducing the chemistry of the nonmetals, it is worth commenting that although they are fewer in number than the metals, they form more compounds. This is because they combine with metals as well as with each other. The four elements examined in this chapter are discussed together because they are the principal building blocks of all living things. Their chemistry is therefore important to everyone.

Objectives

When students have completed this chapter, they should be able to:

Comment on the relative abundance of hydrogen in the universe and on earth.

Name and describe the properties of the isotopes of hydrogen.

Give industrial methods for preparing hydrogen and describe its uses.

Describe how hydrogen can be prepared in the laboratory.

Give chemical equations that illustrate the formation and reactions of saltlike hydrides.

Define catenation and give examples.

Describe in detail the two general methods of preparing nonmetal hydrides discussed here.

Give the pros and cons of a hydrogen economy.

Give the sources of carbon and describe the properties of the free element.

Describe the properties and reactions of the oxides and oxoacids of carbon.

Discuss the preparation and properties of covalent and ionic carbides, cyanides, and carbon disulfide.

Describe the sources, preparation, and commercial uses of oxygen.

Describe how O_2 can be prepared in the laboratory.

Describe the properties of ozone, and how it is involved in photochemical smog and the ozone layer in the upper atmosphere.

Describe the preparation and properties of compounds of oxygen.

Describe the sources, preparation, and uses of nitrogen.

Discuss the preparation and reactions of ionic nitrides.

Discuss the preparation, properties, structures, and reactions of covalent compounds of nitrogen ranging from oxidation states of 3- to 5+, especially the Haber Process and the Ostwald Process.

Describe the reactions that occur during a typical photochemical smog episode.

Answers to Questions

19.1 Hydrogen

19.2 The Earth's gravity wasn't strong enough to hold onto the hydrogen.

19.3 It reacts so readily with oxygen.

19.4 Advantage: less dense than He, so it has greater lifting power. Disadvantage: hydrogen burns but helium doesn't.

19.5 Because there are no electrons below hydrogen's valence shell. It is in Group IA because of its valence shell configuration.

19.6 ^1_1H, protium; ^2_1H, deuterium; ^3_1H, tritium. Tritium is radioactive and is dangerous because it can replace ordinary hydrogen in molecules in the body.

19.7 (a) $CH_4 + H_2O \longrightarrow CO + 3H_2$

$CO + H_2O \longrightarrow CO_2 + H_2$

(b) $C + H_2O \longrightarrow CO + H_2$

19.8 $2CO + O_2 \longrightarrow 2CO_2$

$2H_2 + O_2 \longrightarrow 2H_2O$

19.9 Hydrogen is a byproduct in the preparation of caustic soda.

19.10 Preparation of ammonia.

19.11 $Zn + H_2SO_4 \longrightarrow ZnSO_4 + H_2$
See Figure 19.1.

19.12 $CO + 2H_2 \xrightarrow[\substack{\text{pressure} \\ \text{heat}}]{\text{catalyst}} CH_3OH$

Provides a route from coal to a liquid fuel.

19.13

$$H-C\equiv C-H \ + \ 2H_2 \longrightarrow \begin{array}{c} H \ H \\ | \ \ | \\ H-C-C-H \\ | \ \ | \\ H \ H \end{array}$$

Hydrogenation of vegetable oils gives solid fats.

19.14 Hydrides. NaH is sodium hydride.
$2Na(s) + H_2(g) \longrightarrow 2NaH(s)$

$2NaH + 2H_2O \longrightarrow 2NaOH + H_2(g)$

19.15 One. It has only one valence electron and needs only one more to fill its valence shell.

19.16 (a) nonlinear (b) pyramidal (c) tetrahedral

19.17 Acid-base

$\underset{\text{base}}{H^-} + \underset{\text{acid}}{H_2O} \longrightarrow \underset{\text{acid}}{H_2} + \underset{\text{base}}{OH^-}$

Redox
H^- is oxidized to give H_2, H in H_2O is reduced.

19.18 Linking together of atoms of the same element to form chains. Carbon catenates most. Tendency to catenate decreases going down a group.

19.19 Some have positive ΔG_f°. Those that can be made by direct combination are CH_4, NH_3, H_2O, H_2S, HF, HCl, HBr.

19.20 X^{n-} becomes stronger from right to left in a period (e.g., C^{4-} is stronger base than F^-), and from bottom to top in a group (e.g., O^{2-} is a stronger base than S^{2-}).

19.21 Advantages: it is clean burning and plentiful.
Disadvantages: difficult to store and must first be extracted from water.

19.22 Hydrocarbons and those compounds derived from hydrocarbons by replacing H with other atoms.

19.23 Heating coal in the absence of air.

19.24 In diamond, C is sp^3 hybridized and gives a 3-dimensional interlocking network of bonds. In graphite, C is sp^2 hybridized and arranged in planar sheets with a delocalized π-electron cloud covering upper and lower surfaces. Sheets are stacked one on another and slide over each other easily.

19.25 Lubricant, making electrodes.

19.26 Carbon formed by heating wood in the absence of air. Its large surface area per unit mass allows it to adsorb large numbers of molecules.

19.27 Burning CH_4 in a limited supply of O_2.

$$CH_4 + O_2 \longrightarrow C + 2H_2O$$

19.28 $:C{\equiv}O:$, $:\ddot{O}{=}C{=}\ddot{O}:$

19.29 $HCO_2H(\ell) \xrightarrow{H_2SO_4} H_2O(\ell) + CO(g)$

19.30 $Fe_2O_3(s) + 3CO(g) \xrightarrow{heat} 2Fe(s) + 3CO_2(g)$

19.31 Compound formed between a metal and CO. $Ni(CO)_4$.

19.32 Commercially: from limestone.

$$CaCO_3 \xrightarrow{\text{heat}} CaO + CO_2$$

Laboratory:

$$CaCO_3(s) + 2H^+(aq) \longrightarrow Ca^{2+}(aq) + CO_2(g) + H_2O$$

19.33 Making Na_2CO_3, refrigeration, beverage carbonation.

19.34 By photosynthesis, to make cellulose (polymer of glucose).

19.35 $CaCO_3(s) + H_2CO_3(aq) \rightleftharpoons Ca(HCO_3)_2(aq)$

19.36 As H_2O evaporates, H_2CO_3 decomposes and reaction in the answer to Question 19.35 is shifted to the left.

19.37 Water containing Ca^{2+}, Mg^{2+}, Fe^{3+}. Addition of washing soda, $Na_2CO_3 \cdot 10\,H_2O$ or heating it, if it contains HCO_3^-.

19.38 Wash with dilute acid, which will dissolve $CaCO_3$.

19.39 SiC. Si replaces half of C atoms in the diamond structure.

$$SiO_2(s) + 3C(s) \longrightarrow 2CO(g) + SiC(s)$$

19.40 $Al_4C_3 + 12H_2O \longrightarrow 4Al(OH)_3 + 3CH_4(g)$

19.41 $CaC_2 + 2H_2O \longrightarrow Ca(OH)_2 + C_2H_2$

19.42 Carbon atoms are located between atoms of host lattice. Tungsten carbide, WC.

19.43 $NH_3 + CH_4 \longrightarrow HCN + 3H_2$

HCN deactivates critical enzymes.

19.44 $:\!\ddot{S}\!=\!C\!=\!\ddot{S}\!:$; CS_2 is very flammable.

19.45 The atmosphere.

19.46 Dihydrogen, dioxygen, dinitrogen.

19.47 Catalytic decomposition of $KClO_3$

$$2KClO_3 \xrightarrow[\text{heat}]{MnO_2} 2KCl + 3O_2$$

19.48 Manufacture of steel.

19.49   (Lewis structure of ozone: central $\ddot{O}$ double-bonded to one $\overset{..}{O}$ and single-bonded to $\overset{..}{O}$)   Made by electric discharge through O_2.

O_3 doesn't form poisonous compounds with impurities in the water, but it does kill bacteria.

19.50 $O_2 \xrightarrow{h\nu} 2O$

$O_2 + O \longrightarrow O_3$

Ozone protects the earth by absorbing harmful UV radiation.
Removal of O_3 from the ozone layer:

$\qquad NO + O_3 \longrightarrow NO_2 + O_2$

$\qquad NO_2 + O \longrightarrow O_2 + NO$

and, with Freons

$\qquad CFCl_3 \xrightarrow{h\nu} CFCl_2 + Cl$

$\qquad Cl + O_3 \longrightarrow ClO + O_2$

$\qquad ClO + O \longrightarrow Cl + O_2$

19.51 $O^{2-} + H_2O \longrightarrow 2OH^-$

19.52 Removal of Fe_2O_3 (rust) by reaction with acid.

19.53 $Al_2O_3 + 6H^+ \longrightarrow 2Al^{3+} + 3H_2O$

$Al_2O_3 + 2OH^- \longrightarrow 2AlO_2^- + H_2O$

19.54 From elemental carbon: $C + O_2 \longrightarrow CO_2$

From lower oxides: $2CO + O_2 \longrightarrow 2CO_2$

From hydride: $CH_4 + 2O_2 \longrightarrow CO_2 + 2H_2O$

19.55 Its ΔG_f° is positive, so $K_c \ll 1$ for its formation.

19.56 $4NH_3 + 5O_2 \longrightarrow 4NO + 6H_2O$

19.57 (a) $Cu + 2NO_3^- + 4H^+ \longrightarrow Cu^{2+} + 2NO_2 + 2H_2O$ (concentrated)

(b) $3Cu + 2NO_3^- + 8H^+ \longrightarrow 3Cu^{2+} + 2NO + 4H_2O$ (dilute)

19.58 (a) $2Na + O_2 \longrightarrow Na_2O_2$ (b) $K + O_2 \longrightarrow KO_2$

19.59 $H-\overset{..}{O}-\overset{..}{O}-H$, see Figure 19.6.

19.60 $2H_2O_2(\ell) \longrightarrow O_2(g) + 2H_2O(\ell)$

19.61 Because of its very strong triple bond.

19.62 Making ammonia; as an unreactive gaseous blanket during the manufacture of chemicals; as a refrigerant (liquid N_2).

19.63 Bacteria that remove N_2 from the air and make nitrogen compounds with it.

19.64 Warming a solution containing NH_4^+ and NO_2^-.

19.65 Lithium

19.66 It hydrolyzes to give NH_3.
$Mg_3N_2 + 6H_2O \longrightarrow 3Mg(OH)_2 + 2NH_3$

19.67 (a) NH_3 (b) NH_2OH (c) N_2O (d) N_2O_3 (e) N_2O_5

19.68 $N_2 + 3H_2 \xrightarrow[\text{catalyst}]{\text{iron}} 2NH_3$
Temp; 400-500°C. Pressure; several hundred atmospheres. These conditions chosen to produce maximum amount of NH_3 in shortest time.

19.69 It forms hydrogen bonds with water.

19.70 NH_3

19.71 KNH_2. In water, NH_2^- hydrolyzes immediately to give NH_3 because NH_2^- is a very strong base.

19.72 $NH_4Cl + CaO \longrightarrow NH_3 + H_2O + CaCl_2$
It can't be collected by displacement of water because it is so soluble in water.

19.73 Warm it with base, which causes NH_3 to be evolved. The NH_3 can be detected by its odor or its basic effect on moist blue litmus paper.

19.74 $4NH_3 + 5O_2 \longrightarrow 4NO + 6H_2O$
$2NO + O_2 \longrightarrow NO_2$

$$3NO_2 + H_2O \longrightarrow 2H^+ + 2NO_3^- + NO$$

19.75   A reaction in which the same chemical undergoes both oxidation and reduction.

19.76   They prolonged WWI by allowing Germany to make munitions without having to import nitrates from Chile.

19.77 $H-\overset{..}{N}-\overset{..}{N}-H$, see Figure 19.7.
        $\quad\quad |\quad\ |$
        $\quad\quad H\ \ H$

19.78 $2NH_3 + NaOCl \longrightarrow N_2H_4 + NaCl + H_2O$

Used in rockets because combustion of hydrazine is very exothermic.

19.79 $H-\overset{..}{N}-\overset{..}{O}-H$ Basicities increase: $NH_2OH < N_2H_4 < NH_3$.
        $\quad\quad |$
        $\quad\quad H$

19.80   They can react to form N_2H_4, which is very poisonous.

19.81 $$NH_4NO_3(\ell) \xrightarrow{\text{heat}} N_2O(g) + 2H_2O(g)$$
        $:\overset{..}{N}=N=\overset{..}{O}: \longleftrightarrow :N\equiv N-\overset{..}{\underset{..}{O}}:$

19.82   The ΔG_f^0 of N_2O, NO, NO_2 are all positive. Their decompositions therefore have negative $\Delta G°$, so they should proceed far toward completion. Stabilities of the oxides results because their decomposition is slow.

19.83   See Figure 19.8 and Figure 5.22.
        Bond order: $NO^- < NO < NO^+$
        $$O_2^{2-} < O_2^- < O_2 < O_2^+$$
        Bond length: $NO^- > NO > NO^+$
        $$O_2^{2-} > O_2^- > O_2 > O_2^+$$
        Bond energy: $NO^- < NO < NO^+$
        $$O_2^{2-} < O_2^- < O_2 < O_2^+$$

19.84 $:\overset{..}{\underset{..}{O}}-N=\overset{..}{O}:$,

$$2NO_2 \rightleftharpoons N_2O_4$$

19.85 $N_2 + O_2 \rightleftharpoons 2NO$

$2NO + O_2 \longrightarrow 2NO_2$

$NO_2 \xrightarrow{h\nu} NO + O$

$O + O_2 \longrightarrow O_3$

$O_3 + hydrocarbons \longrightarrow PAN$

19.86 NO_2, PAN is

$$\overset{\displaystyle :O:}{\underset{\displaystyle}{R-C-\overset{..}{\underset{..}{O}}-\overset{..}{\underset{..}{O}}-NO_2}}$$

19.87

$$\overset{..}{:O} = \overset{}{N} - \overset{}{N} \overset{\displaystyle \overset{..}{O}:}{\underset{\displaystyle \overset{..}{O}:}{}}$$

$NO + NO_2 \xrightarrow{cool} N_2O_3$

N_2O_3 is anhydride of HNO_2.

19.88 $3HNO_2 \longrightarrow HNO_3 + H_2O + 2NO$

19.89

$$\overset{\displaystyle \cdot N}{\underset{\displaystyle :\overset{..}{O} \cdot \quad \cdot \overset{..}{O}:}{}} \qquad \left[\overset{\displaystyle \overset{..}{N}}{\underset{\displaystyle :\overset{..}{O} \cdot \quad \overset{..}{O}:}{}} \right]^{-}$$

NO_2 should have the larger O–N–O bond angle because the single electron on the nitrogen in NO_2 doesn't offer as much resistance to increasing the O–N–O bond angle as does the pair of electrons on N in NO_2^-.

19.90 (a) $H^+ + 5HNO_2 + MnO_4^- \longrightarrow 5NO_3^- + 2Mn^{2+} + 3H_2O$

(b) $2H^+ + 2HNO_2 + 2I^- \longrightarrow I_2 + 2NO + 2H_2O$

19.91 $NaNO_3 + C \xrightarrow{heat} NaNO_2 + CO$

$NaNO_2$ retards growth of harmful bacteria and retains color of the meat.
$NaNO_2$ may also cause cancer if HNO_2 produced in the stomach forms nitroso-animes with proteins, but the risk of food poisoning without the use of $NaNO_3$ may outweigh the risk of cancer if $NaNO_2$ is present in the meat.

19.92

$$\overset{\displaystyle :\overset{..}{O}}{\underset{\displaystyle :\overset{..}{O}:}{}} N - \overset{..}{\underset{..}{O}} - N \overset{\displaystyle :\overset{..}{O}:}{\underset{\displaystyle :\overset{..}{O}:}{}} \qquad (vapor); \quad NO_2^+ NO_3^- \text{ in solid}$$

19.93 Dehydration of HNO_3.

19.94 $N_2O_5 + H_2O \longrightarrow 2HNO_3$

19.95 Photodecomposition gives NO_2.

$4HNO_3 \xrightarrow{h\nu} 4NO_2 + O_2 + 2H_2O$

19.96 $NaNO_3 + H_2SO_4 \xrightarrow{heat} NaHSO_4 + HNO_3$

19.97 Manufacture of fertilizers, explosives, and as a meat preservative.

19.98 NH_4^+

19.99 H^+ is not a strong enough oxidizing agent to dissolve Ag or Cu. Nitric acid contains NO_3^- as an oxidizing agent, which is able to oxidize Ag and Cu.

19.100 One part HNO_3 and three parts HCl, by volume. Chloride ion acts as a complex ion-forming agent to help dissolve the noble metals.

CHAPTER 20

THE CHEMISTRY OF SELECTED NONMETALS, PART II:

PHOSPHORUS, SULFUR, THE HALOGENS, THE NOBLE GASES, AND SILICON

Rationale

In this chapter we complete our discussion of the principal nonmetals. As in earlier chapters, we have not attempted to cover everything, but instead have been content to emphasize the important reactions and compounds of these elements and how they relate to students' lives on a day-to-day basis.

Objectives

When students have completed this chapter, they should be able to:

Describe the sources of phosphorus and how the element is recovered from phosphate rock.

Describe the characteristics of the three allotropes of phosphorus.

Describe the structures of P_4O_6 and P_4O_{10}, how these oxides are prepared, and how they react with water.

Discuss the relationship between phosphoric acid and its various condensed, polymeric forms.

Describe the structures of H_3PO_4, $H_4P_2O_7$, and H_3PO_3.

Discuss the preparation and uses of PCl_3 and PCl_5.

Describe the mining of sulfur by the Frasch process.

Explain the changes that take place when sulfur is heated gradually to its boiling point.

Compare the allotropes of sulfur.

Discuss the preparation and reactions of SO_2, H_2SO_3, SO_3, and H_2SO_4.

Describe the properties and origin of acid rain.

Discuss reactions of H_2S, thioacetamide, and thiosulfate ion.

Give the sources of the halogens and examples of their uses.

Compare the abilities of the halogens to function as oxidizing agents.

Describe the preparation and properties of the hydrogen halides.

Write formulas for the oxoacids of the halogens and describe how those of chlorine are prepared.

Discuss the effects of size in determining the number of halogen atoms that a nonmetal can bond to in its halogen compounds.

Discuss the interplay of kinetics and thermodynamics in the hydrolysis of non-metal halides and the disproportionation of the hypohalites.

Discuss the structures of and bonding in noble gas compounds.

Describe the production and refining of silicon.

Discuss the structures of the silicates from the point of view of increasing complexity that accompanies condensation.

Discuss the synthesis and properties of silicones.

Answers to Questions

20.1 $Ca_3(PO_4)_2$

20.2   In DNA, phospholipids (cell membranes), energy storage.

20.3   (a) $1s^2 2s^2 2p^6 3s^2 3p^3$ (d) $1s^2 2s^2 2p^6 3s^2 3p^5$

      (b) $1s^2 2s^2 2p^6 3s^2 3p^4$ (e) $1s^2 2s^2 2p^6 3s^2 3p^2$

      (c) $1s^2 2s^2 2p^5$

20.4   Tetrahedral (see Page 622). The P—P—P bonds are strained and quite weak.

20.5 $2Ca_3(PO_4)_2 + 6SiO_2 + 10C \longrightarrow 6CaSiO_3 + 10CO + P_4$

20.6   A furnace in which the contents are heated by the passage of an electric current through them.

20.7   Red phosphorus - believed to consist of P_4 tetrahedra joined to each other. Black phosphorus - layers of phosphorus atoms in which atoms in a given layer are covalently bonded to each other. Binding between layers is weak. Both of these forms are less reactive than white phosphorus.

20.8 $12Na + P_4 \longrightarrow 4Na_3P$

      $Na_3P + 3H_2O \longrightarrow 3NaOH + PH_3$

20.9   (a) $P_4 + 5O_2 \longrightarrow P_4O_{10}$ (b) $P_4 + 3O_2 \longrightarrow P_4O_6$

20.10 (a) See Figure 20.1.   (b) See Figure 20.2.

20.11 (a) $P_4O_{10} + 6H_2O \longrightarrow 4H_3PO_4$ (b) $P_4O_6 + 6H_2O \longrightarrow 4H_3PO_3$

20.12 A substance that removes H_2O from a gas mixture. P_4O_{10} reacts with water to give H_3PO_4.

20.13 $Ca_3(PO_4)_2 + 3H_2SO_4 + 6H_2O \longrightarrow 3CaSO_4 \cdot 2H_2O + 2H_3PO_4$

20.14 Concentrated H_3PO_4.

20.15 Combustion of phosphorus, reaction of P_4O_{10} with H_2O.

20.16 Fertilizers, food additives, detergents.

20.17 $Mg(H_2PO_4)_2$ magnesium dihydrogen phosphate

      $MgHPO_4$ magnesium hydrogen phosphate

      $Mg_3(PO_4)_2$ magnesium phosphate

20.18 H_3PO_4

20.19 $H_2PO_4^-$ and HPO_4^{2-}. $H_2PO_4^-$ neutralizes strong base, HPO_4^{2-} neutralizes strong acid.

20.20 Water softener, cleaning agent. Hydrolysis of PO_4^{3-} makes the solution basic.

130

20.21

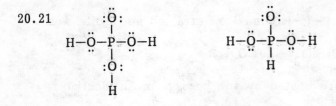

20.22 A mixture of calcium sulfate and calcium dihydrogen phosphate.
$Ca_3(PO_4)_2 + 2H_2SO_4 + 4H_2O \longrightarrow 2CaSO_4 \cdot 2H_2O + Ca(H_2PO_4)_2$

20.23 $Ca_3(PO_4)_2$ is insoluble, so little phosphate enters solution to be available for absorption by plants.

20.24

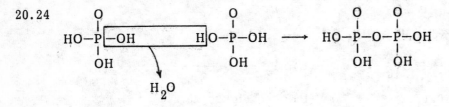

20.25 A PO_4 tetrahedron.

20.26 PO_3^- is metaphosphate. Formed by condensation of $H_2PO_4^-$.

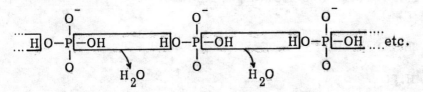

20.27 See Page 627. It is used in solid detergents.

20.28 3 mol NaH_2PO_4 to 2 mol $NaHPO_4$ (the HPO_4^- units terminate the ends of the chains).

20.29 They promote algae blooms which deplete the oxygen from the lake when the algae die and decompose. This kills fish and other aquatic life.

20.30 Phosphorous acid.
$Mg(H_2PO_3)$ and $MgHPO_3$ (H_3PO_3 is a diprotic acid).

20.31 H_3PO_4 is a poor oxidizing agent. H_3PO_3 is a moderately good reducing agent.

20.32 See Figure 20.3. PCl_5 exists as $PCl_4^+PCl_6^-$ in the solid. In PCl_3, phosphorus uses sp^3 hybrids; in PCl_5 it uses dsp^3 hybrids.

20.33 Elemental sulfur. It means "stone that burns."

20.34 As deposits of elemental sulfur, in sulfates, in sulfides.

20.35 Rhombic sulfur and monoclinic sulfur. They have different packing of S_8 rings in their crystals. When heated, solid sulfur melts to give amber liquid containing S_8 rings. The rings break and join with S_x chains as the liquid darkens and thickens. The S_x chains break into smaller fragments at still higher temperatures.

20.36 A form of sulfur produced when hot sulfur containing long S_x chains is suddenly cooled.

20.37 Superheated water is pumped into the sulfur deposit where it melts the sulfur. This is then foamed to the surface with compressed air.

20.38 SO_2

20.39 $S + O_2 \longrightarrow SO_2$

$2SO_2 + O_2 \xrightarrow{\text{catalyst}} 2SO_3$

$SO_3 + H_2SO_4 \longrightarrow H_2S_2O_7$

$H_2S_2O_7 + H_2O \longrightarrow 2H_2SO_4$

20.40 The reaction of SO_2 with O_2 is slow.

20.41 By reaction of a sulfite (e.g., Na_2SO_3) or bisulfite (e.g., $NaHSO_3$) with an acid.

20.42 (a) $SO_2 + H_2O \longrightarrow H_2SO_3$ (b) $SO_3 + H_2O \longrightarrow H_2SO_4$

20.43 Rain falling through air that is polluted with SO_2 becomes acidic because SO_2 reacts with water to form H_2SO_3. It causes structural damage to buildings, causes corrosion of metals, and kills fish and plants.

20.44 H_2SO_4 is a stronger acid than H_2SO_3. The two lone oxygens on S in H_2SO_4 cause a greater polarization of the O—H bonds than does the single lone oxygen attached to S in H_2SO_3.

20.45 Production of fertilizers, refining petroleum, lead storage batteries, manufacture of other chemicals, steel industry.

132

20.46 $\ddot{O}=\ddot{S}-\ddot{O}:$ $\longleftrightarrow$ $:\ddot{O}-\ddot{S}=\ddot{O}$, nonlinear

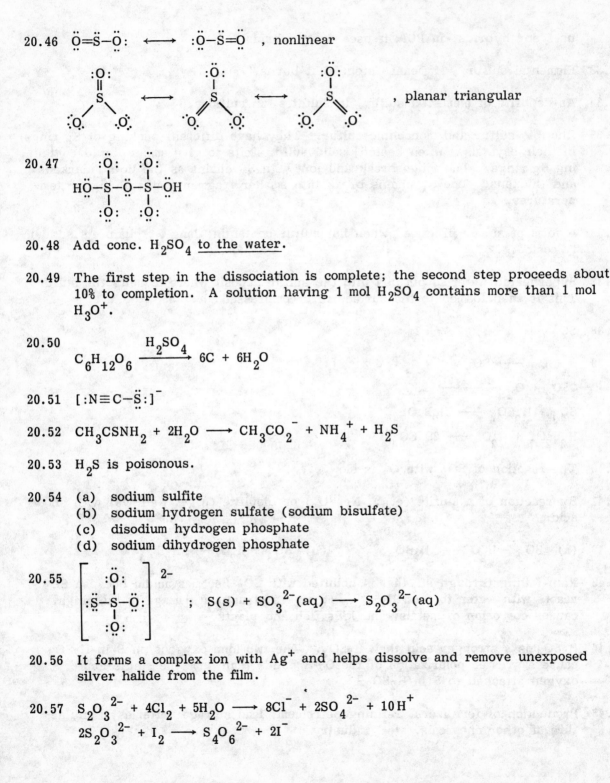

, planar triangular

20.47 $HO-S-O-S-OH$ (with O substituents)

20.48 Add conc. H_2SO_4 <u>to the water.</u>

20.49 The first step in the dissociation is complete; the second step proceeds about 10% to completion. A solution having 1 mol H_2SO_4 contains more than 1 mol H_3O^+.

20.50 $C_6H_{12}O_6 \xrightarrow{H_2SO_4} 6C + 6H_2O$

20.51 $[:N\equiv C-\ddot{S}:]^-$

20.52 $CH_3CSNH_2 + 2H_2O \longrightarrow CH_3CO_2^- + NH_4^+ + H_2S$

20.53 H_2S is poisonous.

20.54 (a) sodium sulfite
       (b) sodium hydrogen sulfate (sodium bisulfate)
       (c) disodium hydrogen phosphate
       (d) sodium dihydrogen phosphate

20.55 $\begin{bmatrix} :\ddot{O}: \\ | \\ :\ddot{S}-S-\ddot{O}: \\ | \\ :\ddot{O}: \end{bmatrix}^{2-}$; $S(s) + SO_3^{2-}(aq) \longrightarrow S_2O_3^{2-}(aq)$

20.56 It forms a complex ion with Ag^+ and helps dissolve and remove unexposed silver halide from the film.

20.57 $S_2O_3^{2-} + 4Cl_2 + 5H_2O \longrightarrow 8Cl^- + 2SO_4^{2-} + 10H^+$
       $2S_2O_3^{2-} + I_2 \longrightarrow S_4O_6^{2-} + 2I^-$

20.58　(a)　I_2 and SO_2　　　　　(d)　Zn^{2+} and H_2

　　　　(b)　Cu^{2+} and SO_2　　　(e)　no reaction

　　　　(c)　S or H_2S and Zn^{2+}

20.59　In the combined state in compounds, normally as halide ions.

20.60　Fluorine:　　CaF_2, Na_3AlF_6, $Ca_5(PO_4)_3F$
　　　　Chlorine:　　NaCl in sea water and mineral deposits.
　　　　Bromine:　　Sea water and brine wells.
　　　　Iodine:　　　Seaweed and as an impurity in saltpeter imported from Chile.

20.61　Electrolysis of HF dissolved in molten KF.

20.62　Lab:　$MnO_2(s) + 4HCl(aq) \longrightarrow MnCl_2(aq) + Cl_2(g) + 2H_2O$

　　　　Commercially:　Electrolysis of molten NaCl or brine.
　　　　Uses:　Treating drinking water, making solvents, manufacture of pesticides.

20.63　F_2:　Pale yellow gas with b.p. of $-188°C$
　　　　Cl_2:　Pale green gas, b.p. = $-34.6°C$
　　　　Br_2:　Dark red liquid, b.p. = $58.8°C$
　　　　I_2:　Dark, metallic-looking solid, m.p. = $113.5°C$

20.64　(a)　$Cl_2 + 2KI \longrightarrow I_2 + 2KCl$　　　(c)　$I_2 + NaCl \longrightarrow$ N.R.

　　　　(b)　$F_2 + 2KBr \longrightarrow Br_2 + 2KF$　　(d)　$Br_2 + 2NaI \longrightarrow I_2 + 2NaBr$

20.65　The F—F bond is quite weak.

20.66　Recovered commercially from sea water by the reaction,
　　　　$2Br^-(aq) + Cl_2(aq) \longrightarrow Br_2(aq) + 2Cl^-(aq)$

　　　　Blowing air through the water removes the volatile Br_2. Uses are in making
　　　　$Pb(C_2H_5)_4$ and AgBr. In the lab, Br_2 can be made by the reaction,
　　　　$MnO_2 + 2Br^- + 4H^+ \longrightarrow Mn^{2+} + Br_2 + 2H_2O$

20.67　Recovered from seaweed and from $NaIO_3$, which is an impurity in Chile
　　　　saltpeter.

20.68　F_2 combines instantly and explosively with H_2 to form HF. Cl_2 combines with
　　　　H_2 explosively if the mixture is heated or exposed to UV light. Reactions of
　　　　H_2 with Br_2 and I_2 are less vigorous.

20.69　$CaF_2 + H_2SO_4 \longrightarrow CaSO_4 + 2HF$

20.70 $NaCl + H_2SO_4 \longrightarrow HCl(g) + NaHSO_4$

HCl is used to remove rust from steel and in the manufacture of other chemicals.

20.71 $NaBr + H_3PO_4 \xrightarrow{\text{heat}} HBr + NaH_2PO_4$

$NaI + H_3PO_4 \xrightarrow{\text{heat}} HI + NaH_2PO_4$

20.72   Volatile SiF_4 is formed. $SiO_2(s) + 4HF(aq) \longrightarrow SiF_4(g) + 2H_2O(\ell)$

20.73   HF > HCl < HBr < HI.   HF is hydrogen bonded into staggered chains (see Page 639).

20.74   (a)   hypobromous acid   (e)   periodic acid
         (b)   sodium hypochlorite   (f)   bromic acid
         (c)   potassium bromate   (g)   sodium iodate
         (d)   magnesium perchlorate   (h)   potassium chlorite

20.75   The same substance undergoes both oxidation and reduction; a portion of it is oxidized while the rest is reduced.

20.76   (a) $Cl_2 + H_2O \rightleftharpoons H^+ + Cl^- + HOCl$

         (b) $Cl_2 + 2OH^- \longrightarrow OCl^- + Cl^- + H_2O$

20.77 $CaCl(OCl) + 2H^+ \longrightarrow Ca^{2+} + H_2O + Cl_2$

20.78 OCl^- is stable, OBr^- reacts moderately fast, OI^- reacts very rapidly. Stability of OCl^- is related to slow rate of reaction rather than to thermodynamic stability.

20.79 $4KClO_3 \xrightarrow{\text{heat}} 3KClO_4 + KCl$

$2KClO_3 \xrightarrow{MnO_2} 2KCl + 3O_2$

20.80   (a)   nonlinear   (e)   T-shaped
         (b)   trigonal bipyramidal   (f)   distorted tetrahedral
         (c)   octahedral   (g)   square pyramidal
         (d)   trigonal pyramidal   (h)   tetrahedral

20.81   Seven fluorine atoms cannot fit around the smaller chlorine atoms.

20.82 Oxygen is too small to accommodate four fluorine atoms and in OF_4 there would be more than an octet of electrons in the valence shell of oxygen, which is not permitted.

20.83 $GeCl_4 + 2H_2O \longrightarrow GeO_2 + 4HCl$

20.84 Si has vacant d orbitals that can be used by attacking H_2O molecules, but C does not.

20.85 NH_3 is a good Lewis base, NF_3 is a poor Lewis base because the highly electro-negative fluorine atoms make the nitrogen a poor electron pair donor.

20.86 Noble gas atoms trapped in cagelike sites in a crystal lattice.

20.87 XeF_4, square planar; XeF_2, linear

20.88 They had believed that the completed octet was chemically inert.

20.89 $SiO_2(s) + 2C(s) \xrightarrow{\text{heat}} Si(s) + 2CO(g)$

20.90 A thin section of a bar of the substance to be refined is melted and the molten zone is gradually moved from one end of the bar to the other. The impurities collect in the molten zone.

20.91 Because Si does not form stable π-bonds to other Si atoms.

20.92
$$
\begin{bmatrix}
& \ddot{\text{O}}: & \\
& | & \\
:\ddot{\text{O}}-\text{Si}-\ddot{\text{O}}: & & \\
& | & \\
& :\ddot{\text{O}}: &
\end{bmatrix}^{4-}
$$

20.93 See Page 649.

20.94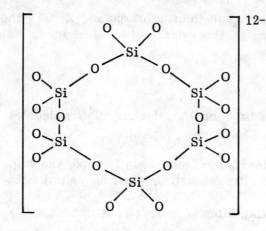

Found in beryl, $Be_3Al_2(Si_6O_{18})$

20.95 See Figure 20.14. Repeating unit is $Si_4O_{11}{}^{6-}$.

20.96 See Figure 20.13.

20.97 Planar sheet silicates - examples are soapstone and talc.

20.98 SiO_2. There are left- and right-handed helices of SiO_4 tetrahedra possible, which gives rise to "left-" and "right-handed" crystals that are mirror images of each other.

20.99 $-Si-O-Si-O-Si-O-$ chain.

20.100

$$
\begin{array}{ccc}
 & CH_3 & CH_3 \\
 & | & | \\
CH_3- & Si-O-Si & -CH_3 \\
 & | & | \\
 & CH_3 & CH_3
\end{array}
$$

CHAPTER 21

THE TRANSITION ELEMENTS

Rationale

This chapter completes a general discussion of the elements by focusing on three aspects of the transition elements: (1) general properties and oxidation states, (2) metallurgy, and (3) coordination compounds. The first three sections examine the properties of the transition elements that set them apart from other members of the periodic table. Section 21.4 (metallurgy) deals with a subject of great industrial and practical significance. Students find this interesting because they encounter daily the products of metallurgical processes. Section 21.5 takes a brief look at the unusual magnetic properties of the iron triad. Next, the physical and chemical properties of the most important transition metals are examined in some detail.

The remainder of the chapter examines one of the most important facets of transition metal chemistry - coordination compounds. In addition to discussing structure, nomenclature, and bonding, you might also mention some of the commonplace applications of coordination chemistry which are described to give students an appreciation of the importance of the subject as it relates to their everyday lives.

Objectives

At the conclusion of this chapter, students should be able to:

Describe similarities and differences between A and B groups of the periodic table.

Distinguish between representative, transition, and inner transition elements.

Write the electronic configurations of the first row transition elements.

Define "lanthanide contraction: and predict its effect on the properties of the transition elements in period 6.

Identify the three steps involved in extracting a metal from its ore.

Describe flotation, roasting, blast furnace, Mond process, Bessemer converter, open hearth process, and basic oxygen process.

Define: amalgam, gangue, slag, cast iron, pig iron, ferromagnetism, domains, ligand, monodentate ligand, polydentate ligand, coordination number, isomerism, enantiomers, racemic, inner orbital complex, outer orbital complex, high spin complex, low spin complex, degenerate energy levels, crystal field splitting energy and crystal field theory.

Explain the phenomenon of ferromagnetism.

Describe the principal chemical and physical properties of chromium, manganese, iron, cobalt, nickel, copper, silver, gold, zinc, cadmium, and mercury.

Name transition metal complexes.

Write formulas of transition metal complexes given their names.

Identify and draw isomers of some transition metal complexes, identifying cis, trans, and optical isomers.

Use valence bond theory to predict the electron configuration of metal complexes.

Use crystal field theory and ligand field theory to predict the paramagnetism or diamagnetism of metal complexes.

Apply crystal field theory to account for the color of transition metal complexes.

Predict and use the crystal field splitting pattern to account for properties of transition metal complexes.

Answers to Questions

21.1   A transition element is one that possesses a partially filled d subshell and fits between Groups IIA and IIIA.

21.2   The inner transition elements possess a partially filled f sublevel that is two below the outermost energy level.

21.3 Compounds of elements in the A and B groups have similar composition, structure, and oxidation states.

21.4 None of these nine elements have counterparts among the representative elements.

21.5 Fe, Co, Ni

21.6 Student answer.

21.7 Four general properties of the transition elements are:
(1) They exhibit multiple oxidation states.
(2) Many transition metal compounds are paramagnetic.
(3) Many compounds are colored.
(4) They tend to form complex ions.

21.8 See Table 21.2.

21.9 Many have a pair of electrons in the outer s orbital. These are the first electrons to be lost.

21.10 Sc, Y, La
$$2M + 6H_2O \longrightarrow 2M(OH)_3 + 3H_2$$

21.11 They have a partially filled d subshell below the outer shell.

21.12 Going from left to right in a period, the lower oxidation states become relatively more stable. Going from top to bottom in a B-group, the higher oxidation states become relatively more stable.

21.13 CrO_4^{2-}

21.14 Ni^{3+}

21.15 Cr^{2+}

21.16 Cu^{3+}

21.17 The differences in electron configuration occur in a subshell that is two shells below the outer shell.

21.18

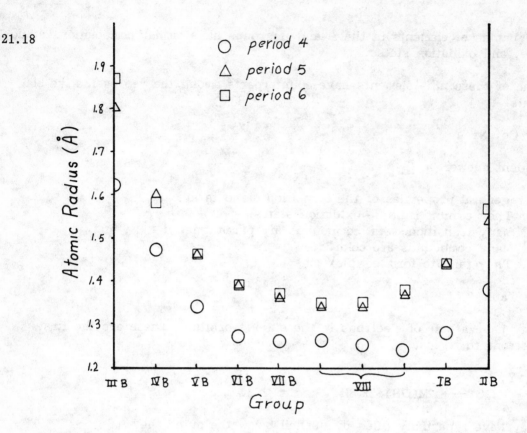

The atomic radii of second and third row elements are nearly identical from Group IVB onward. In general, atoms with large numbers of unpaired electrons have small atomic radii.

21.19 Because of the lanthanide contraction, Hf is very nearly the same size as Zr. Thus their chemical properties are very similar.

21.20 An ore is a mixture of substances that contains a particularly desirable constituent in a high enough concentration that its extraction from the mixture is economically worthwhile.

21.21 The three steps are: concentration, reduction, and refining.

21.22 Gold has a much larger density than sand and mud particles and is not as easily washed away.

21.23 An amalgam is a solution of a metal in mercury. Gold ore is mixed with mercury, which dissolves the metallic gold. The mercury is then separated from the stone and distilled, leaving the pure gold behind.

21.24 In flotation, the ore is finely ground and added to a mixture of oil and water. A stream of air is then blown through the mixture and the oil covered mineral is carried to the surface where it can be removed. In roasting, a sulfide ore is heated in air, converting the metal sulfide to an oxide that is more conveniently reduced.

21.25 $Al_2O_3 + 2OH^- \longrightarrow 2AlO_2^- + H_2O$

$AlO_2^- + H_2O + H^+ \longrightarrow Al(OH)_3$

$2Al(OH)_3 \longrightarrow Al_2O_3 + 3H_2O$

21.26 Because of low toxicity and because it doesn't darken in the presence of H_2S. Titanium is considerably less dense than steel and does not lose its strength at high temperature.

21.27 (a) $3Fe_2O_3 + CO \longrightarrow 2Fe_3O_4 + CO_2$

$Fe_3O_4 + CO \longrightarrow 3FeO + CO_2$

$FeO + CO \longrightarrow Fe + CO_2$

(b) $CaCO_3 \xrightarrow{\text{heat}} CaO + CO_2$

$CaO + SiO_2 \longrightarrow CaSiO_3$

21.28 To remove impurities and to lower the carbon content. The Bessemer converter, open hearth furnace and the basic oxygen process are described at the end of Section 21.4 (Pages 667-668). The basic oxygen process is the principal steel-making method today.

21.29 The Mond process is used in the refining of nickel. In this process impure nickel is treated with carbon monoxide at moderately low temperatures. The $Ni(CO)_4$ gas that is produced is then heated to 200°C and decomposes to give pure nickel plus CO.

21.30 Paramagnetic substances are weakly drawn into a magnetic field whereas ferromagnetic substances are strongly drawn into a magnetic field. Both properties are a result of unpaired electrons. When a substance that possesses ferromagnetism is melted, it becomes simply paramagnetic because the domains are lost. A ferromagnetic substance can become permanently magnetized by placing it in a strong magnetic field so that the domains become aligned with the field. The domains remain aligned even after removal of the external field.

21.31 Because it is lustrous and resistant to corrosion. Oxidation states are 2+, 3+, and 6+.

21.32 Stainless steel is very resistant to corrosion. It usually is not ferromagnetic.

21.33 $Cr(H_2O)_6^{3+}$ is a weak acid. $Cr(H_2O)_6^{3+} + H_2O \quad Cr(H_2O)_5OH^{2+} + H_3O^+$

21.34 $Cr(H_2O)_3(OH)_3 + H_3O^+ \longrightarrow Cr(H_2O)_4(OH)_2 + H_2O$

$Cr(H_2O)_3(OH)_3 + OH^- \longrightarrow Cr(H_2O)_2(OH)_4 + H_2O$

21.35 $CrO_2(OH)_2$ contains two lone oxygens. Because of their high electronegativity, electron density is drained from the O—H bonds, making it easy for the hydrogen to be removed as H^+. This is not the case, however, for $Cr(OH)_3$.

21.36 Oxides of metals in high oxidation states are anhydrides of oxoacids that have one or more lone oxygens and are therefore acidic.

21.37

21.38 SO_4^{2-} and $S_2O_7^{2-}$

21.39 2+, 4+, 6+, 7+

21.40 It is deeply colored and its reduction product in acid solution, Mn^{2+}, is nearly colorless. In neutral or basic solution, the reduction product is brown, insoluble MnO_2.

21.41 $3MnO_4^{2-} + 4H^+ \longrightarrow 2MnO_4^- + MnO_2 + 2H_2O$

21.42 It is abundant and easily extracted from its ores.

21.43 FeO, Fe_2O_3, Fe_3O_4. Only Fe_3O_4 is magnetic.

21.44 $Fe \longrightarrow Fe^{2+} + 2e^-$

$2e^- + \frac{1}{2}O_2 + H_2O \longrightarrow 2OH^-$

$Fe^{2+} + 2OH^- \longrightarrow Fe(OH)_2$

$4Fe(OH)_2 + O_2 + 2H_2O \longrightarrow 4Fe(OH)_3$

21.45 $2HCl + Fe \longrightarrow FeCl_2 + H_2$

$2HCl + Mn \longrightarrow MnCl_2 + H_2$

21.46 It is useful in high-temperature alloys that are employed in tools for cutting and machining other metals at high speed.

21.47 It is resistant to corrosion and its alloys are resistant to impact.

21.48 Because of the ion, $Ni(H_2O)_6^{2+}$, which is green. NiO_2 is used as the cathode in nickel-cadmium batteries.

21.49 Their reactivity decreases from Cu to Ag to Au, and their occurrence in nature as the free metal increases from Cu to Ag to Au.

21.50 Copper - electrical wire

Silver - photography

Gold - plating of low voltage electrical contacts

21.51 Principal oxidation states: copper, 1+ and 2+; silver, 1+; gold, 1+ and 3+.

21.52 H^+ is not a strong enough oxidizing agent. Copper and silver react with HNO_3.

$3Cu + 8HNO_3 \longrightarrow 3Cu(NO_3)_2 + 2NO + 4H_2O$

$(3Cu + 8H^+ + 2NO_3^- \longrightarrow 3Cu^{2+} + 2NO + 4H_2O)$

$3Ag + 4HNO_3 \longrightarrow 3AgNO_3 + NO + 2H_2O$

$(3Ag + 4H^+ + NO_3^- \longrightarrow 3Ag^+ + NO + 2H_2O)$

21.53 $Au + 6H^+ + 3NO_3^- + 4Cl^- \longrightarrow AuCl_4^- + 3H_2O + 3NO_2$

21.54 AgCl, AgBr, AgI

21.55 Add HCl to the solution suspected to contain Ag^+. If a precipitate forms, separate it from the solution and add aqueous ammonia. Then acidify the ammonia solution. A white precipitate of AgCl confirms the presence of Ag^+ in the original solution.

$Ag^+ + Cl^- \longrightarrow AgCl(s)$

$AgCl(s) + 2NH_3 \longrightarrow Ag(NH_3)_2^+ + Cl^-$

$Ag(NH_3)_2^+ + Cl^- + 2H^+ \longrightarrow AgCl(s) + 2NH_4^+$

21.56 The deep blue $Cu(NH_3)_4^{2+}$ ion is formed.

21.57 $Zn(s) + H_2SO_4(aq) \longrightarrow ZnSO_4(aq) + H_2$

$Cd(s) + H_2SO_4(aq) \longrightarrow CdSO_4(aq) + H_2$

$Hg(\ell) + H_2SO_4(aq) \longrightarrow$ no reaction

21.58 The protection of a metal from corrosion by connecting it electrically to a metal that is more easily oxidized and which will be preferentially oxidized.

21.59 Covering a metal with a coating of zinc. It protects steel by providing a barrier to oxygen and moisture and by cathodic protection.

21.60 Cadmium is used when a basic environment is anticipated because zinc is attacked by base but cadmium is not. Cadmium is not used more often because it is less abundant than zinc and because its salts are very toxic.

21.61 Brass and bronze.

21.62 Paint pigment, sun screen, fast-setting dental cements.

21.63 $HgCl_2$ is a weak electrolyte. $HgCl_2 + H_2O \rightleftharpoons Hg(OH)Cl + H^+ + Cl^-$

21.64 Add HCl, which will cause Hg_2Cl_2 to precipitate. Treatment of Hg_2Cl_2 with aqueous NH_3 gives a black precipitate because of disproportionation.

$Hg_2^{2+} + 2Cl^- \longrightarrow Hg_2Cl_2$

$Hg_2Cl_2 + 2NH_3 \longrightarrow Hg + Hg(NH_2)Cl + NH_4^+ + Cl^-$

21.65 Ligands tend to be Lewis bases that have an unshared pair of electrons which can be shared with a metal cation. Ligands attached to the metal are considered to be in a first coordination sphere. Compounds of the transition elements in which the metal cation is attached to one or more ions or molecules are called coordination compounds. Ligands which have one atom that can bond to a metal cation are called monodentate ligands. Ligands which have more than one donor atom that can bond to a metal cation are called polydentate ligands. A chelate is formed by a polydentate ligand that can hold the metal cation in its "claws." The coordination number refers to the total number of ligand atoms that are bound to a given metal ion in a complex.

21.66 Removal of silver salts from photographic film, water softening, as a catalyst and to alleviate metal poison.

21.67 EDTA is an antidote in lead poisoning and a water softener in shampoos. It also ties up metal ions that catalyze oxidation of food products and increases the storage life of whole blood.

21.68   See Figure 21.9, Page 681.

21.69

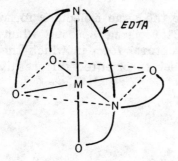

21.70   NTA can coordinate to four sites in an octahedral complex in much the same manner as EDTA.

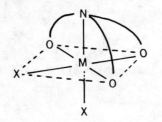

The remaining two sites, X, can be occupied by other ligands or by H_2O molecules.

The NTA would increase the solubility of metal salts by shifting equilibria such as: $MX_n(s) + NTA \rightleftharpoons M(NTA)(aq) + X(aq)$ to the right. This was discussed quantitatively in Chapter 16.

21.71   (a)   hexaamminenickel(II) ion
        (b)   triamminetrichlorochromium(III) ion
        (c)   hexanitrocobaltate(III) ion
        (d)   trioxalatomanganate(III) ion
        (e)   tetraoxomanganate(VII) ion

21.72   (a)   diiodoargentate(I) ion
        (b)   pentaamminechlorochromium(III) ion
        (c)   diamminetetraaquacobalt(II) chloride
        (d)   diaquabis(ethylenediamine)cobalt(III) sulfate
        (e)   tetraamminedichlorochromium(III) chloride

21.73   (a) $[Fe(H_2O)_4(CN)_2]^+$ (c) $K_3[Mn(CN)_6]$ (e) $[CrO_4]^{2-}$
        (b) $[Ni(NH_3)_4(C_2O_4)]$ (d) $[CuCl_4]^{2-}$

21.74 (a) $[AuCl_4]^-$ (d) $[Fe(EDTA)]^{2-}$

      (b) $[Fe(en)_2(NO)_2]_2SO_4$ (e) $[Ag(S_2O_3)_2]^{3-}$

      (c) $[Co(NH_3)_4CO_3]NO_3$

21.75 Isomers are two different compounds that have the same molecular formula but differ in the way their atoms are arranged. Stereoisomers result when a given molecule or ion can exist in more than one structural form in which the same atoms are bound to one another but find themselves oriented differently in space.

21.76 Isomers of $[Co(NH_3)_2Cl_4]^-$

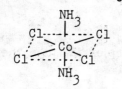

             trans   cis

      Isomers of $[Co(NH_3)_3Cl_3]$

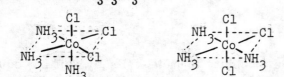

21.77 One isomer cannot be superimposable on its mirror image.

21.78 They or their solutions rotate the plane of polarized light which passes through them.

21.79 Isomers of $[Cr(en)_2Cl_2]^+$

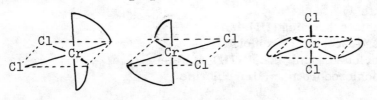

                    cis - (dl pair)   trans

21.80

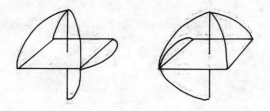

21.81 Enantiomers of a complex are two nonidentical mirror image isomers; e.g. [Co(en)₃]³⁺, Figure 21.14. A <u>racemic</u> compound contains an equal mixture of two enantiomers.

21.82 In an <u>inner orbital</u> complex the transition element's 3d orbitals are being used in hybridization and in an <u>outer orbital</u> complex the 4d orbitals are being used.

21.83
	3d	4s	4p	4d

(a) inner

  ↑ ↑ __ xx xx   xx   xx xx xx   __ __ __ __ __

(b) outer

  ↓↑ ↓↑ ↓↑ ↑ ↑   xx   xx xx xx   xx xx __ __ __

(c) outer

  ↑ ↑ ↑ ↑ ↑   xx   xx xx xx   xx xx __ __ __

(d) inner

  ↓↑ ↓↑ ↓↑ xx xx   xx   xx xx xx   __ __ __ __ __

(e) inner

  ↑ ↑ ↑ xx xx   xx   xx xx xx   __ __ __ __ __

An inner orbital with three unpaired electrons.

21.84 (a) four (outer orbital, high spin)   (b) two (inner orbital, low spin)

21.85 paramagnetic with one unpaired electron

21.86 See Figure 21.20.

148

21.87

$\uparrow$ $\uparrow$ $\uparrow$ $\uparrow$ ___

d orbitals
in free ion

$\overline{d_{z^2}}$ $\overline{d_{x^2-y^2}}$

$\overline{d_{xz}}$ $\overline{d_{yz}}$ $\overline{d_{zy}}$

21.88   A portion of visible light is absorbed in promoting electrons from the t_{2g} level to the e_g level in octahedral complexes. The observed color of the complex is due to the wavelengths that are <u>not</u> absorbed.

21.89   If $\Delta < P$, a paramagnetic complex will be formed; when $\Delta > P$, a diamagnetic complex will form.

21.90   A high-spin complex is one that possesses the maximum number of possible un-paired electrons. A low-spin complex possesses electrons paired in low energy orbitals. Inner orbital complexes would be low-spin and outer orbital complexes would be high-spin.

21.91   See Figure 21.27.

21.92   High Spin   Low Spin

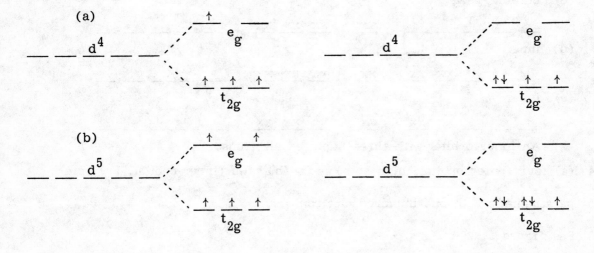

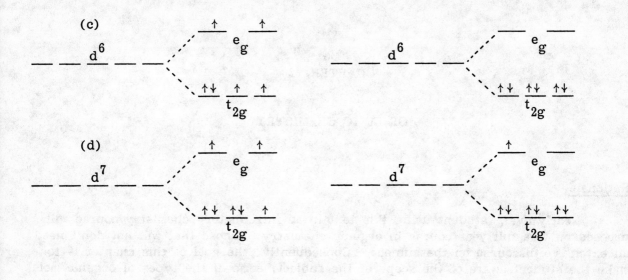

CHAPTER 22

ORGANIC CHEMISTRY

Rationale

Nearly every student using this text in an introductory chemistry course will proceed on to a full year course in organic chemistry. Those that will not don't need an extensive discourse on the subject. Consequently, the goal of this chapter is to make the student aware of the scope of the subject, some of the types of organic molecules that exist, their general properties, and some common everyday uses to which they are put.

The chapter has been written in such a way that it could be moved to an earlier part in the course, if so desired.

Objectives

After completing this chapter, students should be able to:

Define: aliphatic hydrocarbon, aromatic hydrocarbon, derivative, structural isomer, optical isomer, olefin, alkane, alkene, alkyne, asymmetric carbon atom and homologous series.

Draw and name the structural isomers of the first seven alkanes and alkenes.

Apply the I.U.P.A.C. rules of nomenclature to naming hydrocarbons.

Draw the structural and molecular formulas of organic compounds given their I.U.P.A.C. names.

Describe the bonding in aromatic hydrocarbons.

Identify functional groups for alkenes, alkynes, alcohols, aldehydes, carboxylic acids, halides, ketones, esters, amines, amides, ethers.

Describe reactions for the preparation of alkyl halides, ethanol, aldehydes, ketones, carboxylic acids, esters, ethers.

Write an equation for the saponification of an ester.

Distinguish among primary, secondary, and tertiary amines and alcohols.

Give examples of applications of hydrocarbons and their derivatives.

Distinguish between addition and condensation polymerization.

Answers to Questions

22.1 Unsaturated hydrocarbons contain double and/or triple bonds between the carbons. Saturated compounds contain only single bonds.

22.2 As the chain length increases the London forces become stronger because each molecule is attracted to others at more points along the chain.

22.3 (a) $C_{30}H_{62}$ (b) $C_{27}H_{54}$ (c) $C_{33}H_{64}$

22.4 (a) $C_{17}H_{36}$ (b) $C_{17}H_{34}$ (c) $C_{17}H_{32}$ (d) $C_{17}H_{30}$ (e) $C_{17}H_{30}$

22.5 An asymmetric carbon atom must be present for optical isomer.

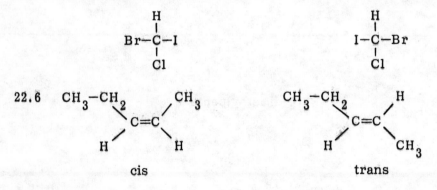

22.6

22.7 Tetrahedral

22.8 Because of the double bond between the carbons. Each carbon uses sp^2 hybrids which gives a planar configuration and there is no rotation about the double bond.

22.9 The chain in butane arises from sp^3 hybridization on the carbon atom. The hybridization is different in two of the carbons in 2-butyne. The middle two carbons are sp hybridized.

22.10 Asterisk indicates asymmetric carbon atom.

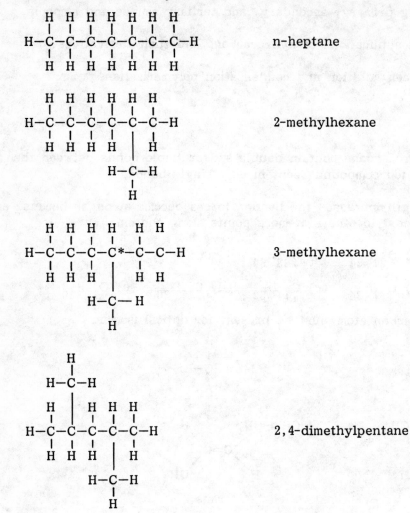

153

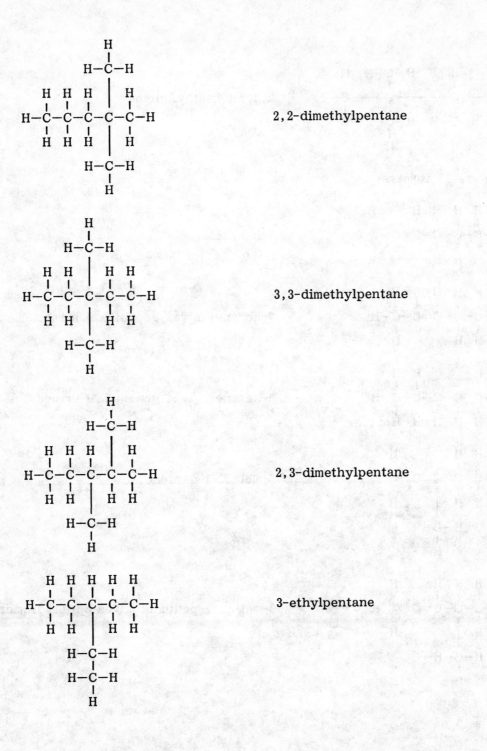

2,2-dimethylpentane

3,3-dimethylpentane

2,3-dimethylpentane

3-ethylpentane

154

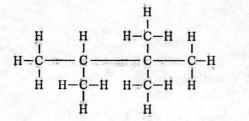

2,2,3-trimethylbutane

22.11   There are 13 isomers:

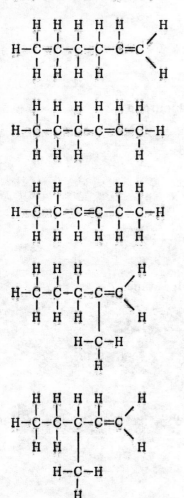

1-hexene

2-hexene; gives geometrical isomer

3-hexene; gives geometrical isomer

2-methyl-1-pentene

3-methyl-1-pentene; gives optical isomers

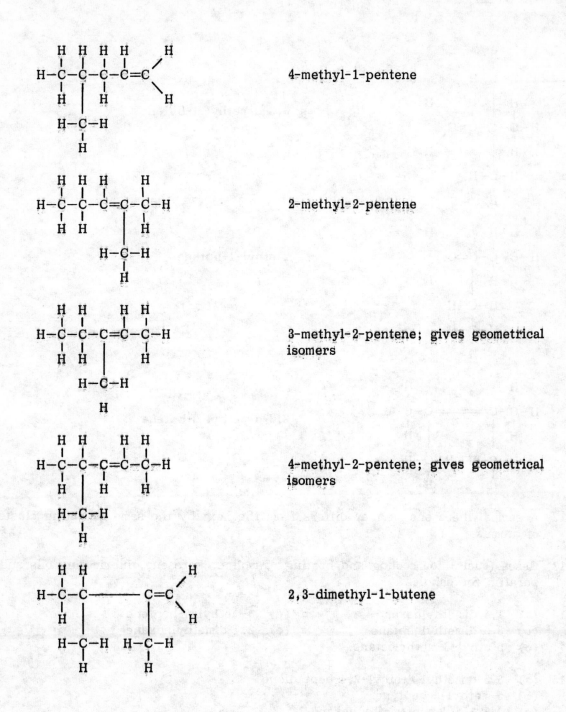

4-methyl-1-pentene

2-methyl-2-pentene

3-methyl-2-pentene; gives geometrical isomers

4-methyl-2-pentene; gives geometrical isomers

2,3-dimethyl-1-butene

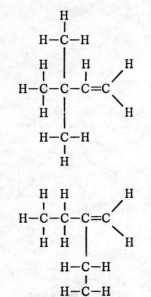

3,3-dimethyl-1-butene

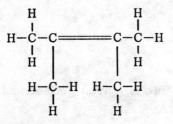

2,ethyl-1-butene

2,3-dimethyl-2-butene

22.12 A series where one member differs from the next by the same repeating cluster of atoms.

22.13 Gases (fuels) for cooling and heating, gasoline, kerosene, lubricating oils and paraffin for candles.

22.14 (a) 2,4-dimethylhexane   (d) 3-methyl-4-heptene
     (b) 3,5-dimethylheptane   (e) 2,4-dimethylhexane
     (c) 5-ethyl-3-methyloctane

22.15 (a) 3,5-dimethyl-4-ethyl-2,4-heptadiene
     (b) 5-methyl-3-heptyne
     (c) 2,3,3,4,4-pentamethylhexane
     (d) 4-methyl-2-pentyne
     (e) 3,4-dimethyl-3,5-octadiene

22.16 This compound is one of the 13 isomers of hexene. These are shown in the answer to Question 22.11.

22.17 (a) CH$_3$–CH–CH$_2$–CH$_2$–CH$_3$
         |
        CH$_3$

    (b) CH$_3$–CH–CH–CH$_3$
          | |
        CH$_3$ CH$_3$

    (c) CH$_3$–C══C—CH$_3$
         | |
        CH$_3$ CH$_3$

    (d) CH$_2$═CH–CH═CH–CH═CH–CH$_2$–CH$_3$

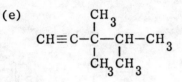

    (e)
             CH$_3$
             |
    CH≡C–C—CH–CH$_3$
           | |
          CH$_3$ CH$_3$

22.18 3,3,4,4,5-pentamethylheptane

22.19

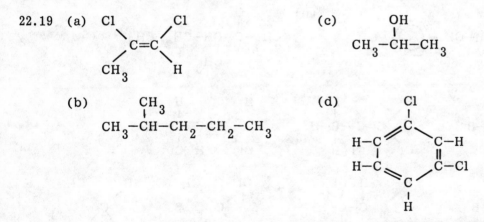

(a) 1,2-dichloro compound (Cl, Cl on C=C with CH$_3$ and H)

(b) CH$_3$–CH–CH$_2$–CH$_2$–CH$_3$ with CH$_3$ branch

(c) CH$_3$–CH–CH$_3$ with OH

(d) dichlorobenzene ring with Cl substituents

22.20 60°, 90°, 108°, 120° (The internal angles in a regular polygon can be calculated from the formula, $(180 - \dfrac{360}{n})$, where n is the number of sides.)

22.21 In the boat form two hydrogen atoms point at each other across the top of the "boat," leading to some repulsion. This is absent in the chair form.

22.22 A carcinogen is a cancer-causing agent. Benzopyrene and graphite are similar because they consist of planes of fused hexagonal rings. Incomplete combustion tends to produce carbon (graphite). Benzopyrene is, in a sense, a "fragment" of graphite.

22.23 The bonding in benzene and graphite are nearly identical Each carbon is sp^2 hybridized which accounts for C—C and C—H bonds, and in addition each has a pure p orbital and electron. These electrons become delocalized in the π orbital around the ring in benzene or across a planar sheet in graphite.

22.24 (a) $CH_3-\overset{\overset{\displaystyle O}{\|}}{C}-H$ (b) $CH_3-\overset{\overset{\displaystyle O}{\|}}{C}-CH_3$ (c) $CH_3-\overset{\overset{\displaystyle O}{\|}}{C}-OH$ (d) CH_3-NH_2

(e) CH_3-CH_2-OH (f) $CH_3-\overset{\overset{\displaystyle O}{\|}}{C}-O-CH_2-CH_3$ (g) CH_3-O-CH_3

22.25 (a) $CH_3-CH_2-\underset{\underset{\displaystyle CH_3}{|}}{C}=CH_2$ (b) $CH_3-\underset{\underset{\displaystyle CH_3}{|}}{\overset{\overset{\displaystyle OH}{|}}{C}}-\underset{\underset{\displaystyle CH_3}{|}}{CH}-CH_3$ (e) [benzene ring with Br, Cl, Cl, Br, Br, Cl substituents]

(c) $CH_3-\underset{\underset{\displaystyle Br}{|}}{CH}-CH_2-$ [benzene ring]   (d) $CH_3-\overset{\overset{\displaystyle O}{\|}}{C}-\underset{\underset{\displaystyle CH_3}{|}}{CH}-CH_2-CH_3$

22.26 [structural isomers drawn:]

Row 1: $\overset{Cl}{\diagdown}\underset{Cl}{\diagup}C=C\underset{H}{\overset{H}{\big|}}-\overset{H}{\underset{H}{\big|}}C-H$; $\overset{H}{\diagdown}\underset{Cl}{\diagup}C=C\underset{Cl}{\overset{H}{\big|}}-\overset{H}{\underset{H}{\big|}}C-H$; $\overset{H}{\diagdown}\underset{Cl}{\diagup}C=C\underset{H}{\overset{H}{\big|}}-\overset{H}{\underset{Cl}{\big|}}C-H$

Row 2: $\overset{H}{\diagdown}\underset{H}{\diagup}C=C\underset{Cl}{\overset{H}{\big|}}-\overset{H}{\underset{Cl}{\big|}}C-H$; $\overset{H}{\diagdown}\underset{H}{\diagup}C=C\underset{Cl}{\overset{H}{\big|}}-\overset{Cl}{\underset{}{\big|}}C-H$; [cyclopropane with Cl, Cl on C and H$_2$C—CH$_2$] ; [cyclopropane with H, Cl on C and H$_2$C—C with Cl, H]

22.27 Alkenes and alkynes tend to undergo addition reactions whereas alkanes tend to undergo substitution reactions.

22.28 Dry cleaning solvents, refrigerants, aerosol propellants, insecticides.

22.29 Student answer.

22.30 (a) cyclohexane   (b) 3-chloro-1-methylbenzene (or 3-chlorotoluene)
      (c) butanone   (d) 1-amino-2-methylpropane

22.31 (a) $CH_3CH_2CH=CH_2$ + HI $\longrightarrow$ $CH_3CH_2CHICH_3$
           1-butene   2-iodobutane

(b) $CH_3CH=CH_2$ + H_2O $\xrightarrow{H_2SO_4}$ $CH_3\underset{\underset{OH}{|}}{C}HCH_3$

         1-propene   2-propanol

(c) $CH_3CH_2CH=C\overset{\displaystyle CH_3}{\underset{\displaystyle CH_3}{}}$ + H_2O $\xrightarrow{H_2SO_4}$ $CH_3CH_2CH_2-\underset{\underset{CH_3}{|}}{\overset{\overset{CH_3}{|}}{C}}-OH$

    2-methyl-2-pentene   2-methyl-2-pentanol

22.32 The molar solubility will decrease because of their large size and because of the decrease in polar character. The molecules become less like H_2O and more like hydrocarbons.

22.33 (a) $CH_3-\overset{\overset{\displaystyle O}{\|}}{C}\diagdown_{OH}$ (b) $CH_3\underset{\underset{CH_3}{|}}{C}H\overset{\overset{\displaystyle O}{}}{C}\diagdown_{OH}$ (c) $CH_3\overset{\overset{\displaystyle O}{\|}}{C}CH_3$

(d) $CH_3CH_2\overset{\overset{\displaystyle O}{}}{C}\diagdown_{OH}$ (e) N.R.   (f) N.R.   (g) N.R.

22.34 (a) mild oxidation; 2-propanol would give acetone, 2-methyl-2-propanol won't be oxidized.
      (b) oxidation: 1-butanol $\longrightarrow$ butyric acid, 2-butanol $\longrightarrow$ ketone
      (c) addition of Br_2 to butene, no reaction with butane
      (d) mild oxidation: ethanal $\longrightarrow$ acetic acid, 2-propanone $\longrightarrow$ N.R.

22.35 C=O, OH and ether.

22.36 (a)
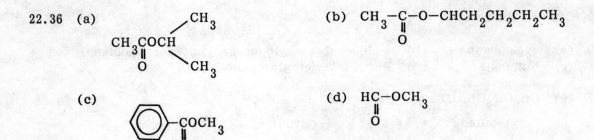
CH$_3$COCH CH$_3$
‖   CH$_3$
O

(b) CH$_3$–C–O–CHCH$_2$CH$_2$CH$_2$CH$_3$
‖
O

(c)
C$_6$H$_5$–COCH$_3$
‖
O

(d) HC–OCH$_3$
‖
O

22.37 (a)

CH$_3$CH$_2$–C–OCH$_3$ + NaOH $\xrightarrow{\text{H}_2\text{O}}$ CH$_3$OH + NaO–C–CH$_2$CH$_3$

(b)

CH$_3$CH$_2$O–CCH$_2$CH$_2$–C–OCH$_2$CH$_3$ + NaOH $\xrightarrow{\text{H}_2\text{O}}$ 2CH$_3$CH$_2$OH +

NaOCCH$_2$CH$_2$CONa

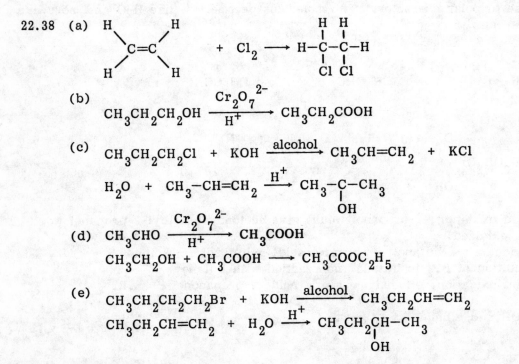

22.38 (a) H$_2$C=CH$_2$ + Cl$_2$ $\longrightarrow$ H–CHCl–CHCl–H

(b) CH$_3$CH$_2$CH$_2$OH $\xrightarrow[\text{H}^+]{\text{Cr}_2\text{O}_7^{2-}}$ CH$_3$CH$_2$COOH

(c) CH$_3$CH$_2$CH$_2$Cl + KOH $\xrightarrow{\text{alcohol}}$ CH$_3$CH=CH$_2$ + KCl

H$_2$O + CH$_3$–CH=CH$_2$ $\xrightarrow{\text{H}^+}$ CH$_3$–C(OH)–CH$_3$

(d) CH$_3$CHO $\xrightarrow[\text{H}^+]{\text{Cr}_2\text{O}_7^{2-}}$ CH$_3$COOH

CH$_3$CH$_2$OH + CH$_3$COOH $\longrightarrow$ CH$_3$COOC$_2$H$_5$

(e) CH$_3$CH$_2$CH$_2$CH$_2$Br + KOH $\xrightarrow{\text{alcohol}}$ CH$_3$CH$_2$CH=CH$_2$

CH$_3$CH$_2$CH=CH$_2$ + H$_2$O $\xrightarrow{\text{H}^+}$ CH$_3$CH$_2$CH(OH)–CH$_3$

$$CH_3CH_2\underset{\overset{|}{OH}}{C}HCH_3 \xrightarrow{\text{oxid}} CH_3-CH_2-\underset{\overset{\|}{O}}{C}-CH_3$$

22.39 (a) $CH_3COOCH_2CH_3$ (b) CH_3CH_2COOH

(c) $CH_3CH_2\underset{\overset{\|}{O}}{C}CH_3$ (d) $CH_3CH_2CH{=}CH_2$ is an acceptable student answer. or $(CH_3CH_2CH_2CH_2)_2O$

22.40
$$CH_3CH_2CH_2O\underset{\overset{\|}{O}}{C}CH_3 + NaOH \xrightarrow{H_2O} CH_3CH_2CH_2OH + NaO\underset{\overset{\|}{O}}{C}CH_3$$

The NaOH drives this reaction by neutralizing the acid as soon as it forms.

22.41 When an aldehyde is reduced with hydrogen, a primary alcohol is produced. When a ketone is reduced in this fashion, a secondary alcohol is produced.

22.42 $(CH_3)_2NH + H_2O \rightleftharpoons (CH_3)_2NH_2^+ + OH^-$

22.43

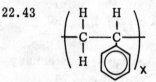

22.44 In addition polymers, monomer units are simply joined together. In condensation polymers this joining together of monomers is done at the expense of a small molecule that is eliminated.

22.45

$$\left(\begin{matrix} & Cl & \\ -CH_2-\underset{\overset{|}{Cl}}{\overset{|}{C}}-CH_2\underset{\overset{|}{Cl}}{CH} \end{matrix}\right)_x \begin{matrix} & Cl & \\ CH_2-\underset{\overset{|}{Cl}}{\overset{|}{C}}-CH_2\underset{\overset{|}{Cl}}{CH}- \end{matrix}$$

22.46 $\left(O-\underset{\overset{\|}{O}}{C}-CH_2-\underset{\overset{\|}{O}}{C}-O-CH_2-CH_2 \right)_x$

22.47 The acid promotes hydrolysis of the nylon.

22.48 Cross-linking is the forming of bonds between adjacent polymer molecules. The greater the degree of cross-linking, the stronger will be the material.

CHAPTER 23

BIOCHEMISTRY

Rationale

Biochemistry is a fascinating subject that never fails to "turn students on." Obviously, the breadth of the subject is vast; so in an introductory course all we can really hope to do is to whet the student's appetite. That is precisely the intent of the chapter. The emphasis is on types of biochemical molecules and their functions in living systems. Because the subject can become so complex, we have deliberately avoided complexity at this stage so that perhaps students might be able to "see the forest through all those trees." In treating this chapter, try to point out simple structural relationships and the way that most biochemical substances are composed of relatively simple building blocks. Emphasize the functions played by the different kinds of biomolecules and the role that their structures have in determining biological activity.

Proteins, with their primary, secondary, tertiary, and (sometimes) quaternary structures, are described first. Key points here are the polymeric nature of these substances and the structural features they exhibit. Enzyme functions are discussed next with emphasis on the "lock-and-key" fit between enzyme and substrate.

The second class of compounds discussed is the carbohydrates. Again, these are described in terms of simple monomers that polymerize to give starch and cellulose.

A third class of biomolecules studied is the lipids. The triglycerides are examined as esters and their saponification products are described. Phospholipids, as components of cellular membranes, and a number of common steroids are also treated.

Finally, the nucleic acids, their polymers (DNA and the RNAs), and DNA replication and protein synthesis are examined. Here again the emphasis is on the polymeric nature of these substances and the way that their structures interact.

Objectives

At the conclusion of this chapter, students should be able to:

Define: α-amino acid; dipeptide; polypeptide; primary, secondary, tertiary, and quaternary structures of a protein; zwitterion; porphyrin; enzyme; enzyme substrate; coenzyme; and competitive inhibition.

Describe in qualitative terms: the α-helix structure, bonds holding the polypeptide in the helical structure, enzyme activity, enzyme inhibition, the contribution of phospholipids to the structure of cellular membranes, the genetic code and the functions of mRNA and tRNA.

Distinguish between: monosaccharides and polysaccharides; starch and cellulose; normal and mutated amino acid sequences.

Define: lipid, soap, steroid, and nucleotide.

Use the concept of base pairing to illustrate DNA replication and the synthesis of proteins through mRNA.

Answers to Questions

23.1   An α-amino acid is a bifunctional organic molecule that contains both a carboxyl and an amine group, with the amine group attached to the carbon adjacent to the carbonyl group.   A peptide with its peptide bond would be

$$
\begin{array}{c}
\quad\ \ H\ \ \ O\quad\ \ H \\
\quad\ \ |\quad\ \ || \quad\ \ | \\
H_2N-C-C-N-C-COOH \\
\quad\ \ |\quad\quad\ \ |\ \ | \\
\quad\ \ R\quad\quad\ H\ \ R
\end{array}
$$

A polypeptide would be

$$
\begin{array}{c}
H\ O\quad\ \ H\ O\quad\ \ H\ O\quad\ \ H\ O \\
|\ \ ||\quad\ \ |\ \ ||\quad\ \ |\ \ ||\quad\ \ |\ \ || \\
-N-C-C-N-C-C-N-C-C-N-C-C- \\
|\ \ |\quad\ \ |\ \ |\quad\ \ |\ \ |\quad\ \ |\ \ | \\
H\ R\quad\ \ H\ R'\quad H\ R''\quad H\ R'''
\end{array}
$$

23.2   The reaction is an acid-base type. The lye can break the peptide bonds by hydrolysis to produce smaller units.

23.3   The primary structure of a protein is the amino acid sequence that exists in the polypeptide. The twisting and turning of the polypeptide chain is its secondary structure. A tertiary structure is the folded chains of the coiled polypeptide. The way in which these folded chains orient themselves with respect to others gives rise to a quaternary structure in substances like hemoglobin.

23.4 They can act as enzymes and catalyze biochemical reactions, or transport substances through an organism. Proteins are the major constituent of such things as muscles, hair, nails, skin and tendons.

23.5 A zwitterion contains one end that is positively charged and one end that is negatively charged. In amino acids the hydrogen on the carboxyl group leaves and attaches itself to the amino group.

23.6 In many proteins the polypeptide chains coil themselves in an α-helix structure (Figure 23.3). Hydrogen bonding between the oxygen in a carbonyl group and the hydrogen attached to a nitrogen that lies in an adjacent loop holds the structure in place.

23.7 In polar solvents the nonpolar R groups are forced toward the center of the folded polypeptide chain and the chain tends to fold in such a way that nonpolar groups do not contact the solvent. Ionic attractions occur between a negatively charged deprotonated carboxyl group and a positively charged protonated amine group.

23.8 The quaternary structure is determined by the way in which the folded proteins orient themselves with respect to one another.

23.9 Hemoglobin carries oxygen in the blood stream whereas myoglobin stores oxygen in muscle tissue. Both contain heme groups.

23.10 The basic square planar structure found in heme is an example of a porphyrin. Two other biologically important porphyrin structures are contained in chlorophyll and vitamin B_{12} coenzyme.

23.11 Enzymes catalyze very specific biochemical reactions to give very nearly 100% product and no by-products. A buildup of by-products would occur if reactions took place without enzymes and this would create a waste disposal problem for the organism.

23.12 An enzyme binds to a reactant molecule, called the substrate. A slight alteration in the shape of the enzyme causes a strain in certain key bonds in the substrate, making them more susceptible to attack. Enzyme inhibition occurs when a substance other than the enzyme substrate becomes bound to the active site of an enzyme, thereby inhibiting its catalytic activity. Sulfa drugs rely on competitive inhibition for their effectiveness. The sulfa drug occupies the active site of an enzyme preventing the production of a critical coenzyme.

23.13 This should not be surprising. In proteins only one isomer of a given amino acid is generally found. This means that isomeric structure is very important in determining physiological activity.

23.14 See p 752. Organophosphate insecticides function by attacking the central nervous system of the insect.

23.15 Glycine is not optically active because it does not have an asymmetric carbon atom.

23.16 Monosaccharides are simple sugars; i.e., they are the simple units that come together to form the more complex polysaccharides.
(a) ribose   (b) glucose

23.17 16

23.18 In starch the polysaccharide chains coil in a helical structure with the polar OH groups pointing outward. Cellulose forms linear chains that interact with each other by hydrogen bonding.

23.19
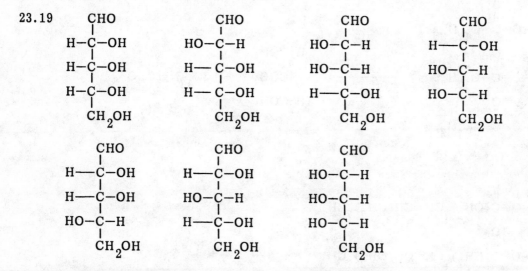

23.20 Glucose and fructose

23.21 Carbohydrates are a source of energy, a source of carbon and a structural element in cells and tissues.

23.22 Lipids are water insoluble substances (e.g., fats and oils) that can be extracted from other cell components by nonpolar organic solvents. Lipids mainly serve as storage of energy-rich fuel for use in metabolism, as components of cell membranes, and as steroids.

23.23 Fatty acids are long unbranched hydrocarbon chains terminated at one end with a carboxyl group.

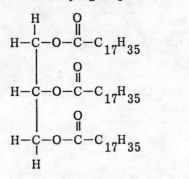

tristearin

Saturated triglycerides tend to be solids and unsaturated ones tend to be liquids.

23.24

$$
\begin{array}{ccc}
\text{H} & \text{O} & \\
| & \| & \\
\text{H}-\text{C}-\text{O}-\text{C}-\text{C}_{17}\text{H}_{35} & & \\
| & \text{O} & \\
& \| & \\
\text{H}-\text{C}-\text{O}-\text{C}-\text{C}_{17}\text{H}_{35} & \xrightarrow[\text{H}_2\text{O}]{\text{OH}^-} &
\begin{array}{c}
\text{H}_2\text{COH} \\
| \\
\text{HCOH} \\
| \\
\text{H}_2\text{COH}
\end{array}
+ \; 3\text{C}_{17}\text{H}_{35}\text{COO}^- \\
| & \text{O} & \\
& \| & \\
\text{H}-\text{C}-\text{O}-\text{C}-\text{C}_{17}\text{H}_{35} & & \\
| & & \\
\text{H} & &
\end{array}
$$

23.25

$$
\begin{array}{cc}
\text{H} & \text{O} \\
| & \| \\
\text{H}-\text{C}-\text{O}-\text{C}(\text{CH}_2)_7\text{CH}=\text{CH}(\text{CH}_2)_7\text{CH}_3 \\
| & \text{O} \\
& \| \\
\text{H}-\text{C}-\text{O}-\text{C}(\text{CH}_2)_7\text{CH}=\text{CH}(\text{CH}_2)_7\text{CH}_3 \\
| & \text{O} \\
& \| \\
\text{H}-\text{C}-\text{O}-\text{C}(\text{CH}_2)_7\text{CH}=\text{CH}(\text{CH}_2)_7\text{CH}_3 \\
| & \\
\text{H} &
\end{array}
$$

The reaction of triolein with H_2 would saturate the molecule, making tristearin which is a solid.

23.26 A soap is an anion that forms when a fat is saponified. In water these anions group themselves into small globules called micelles. The polar head dissolves in water and the nonpolar tail dissolves the grease.

23.27 In phospholipids only two fatty acid molecules are esterified to glycerol. The third position is esterified to a phosphoric acid group which in turn is esterified to another alcohol.

23.28 Phospholipids in membranes appear to be arranged on a bilayer in which the nonpolar tails face each other, exposing the polar heads to the aqueous environment on either side of the membrane.

23.29 Steroids possess the fused ring structure shown on Page 761.

23.30 The monomer units that make up the nucleic acids are called nucleotides. In RNA the five-carbon sugar that makes up the nucleotide is ribose whereas the pentose in DNA is deoxyribose.

23.31

Ⓣ Ⓖ Ⓐ Ⓖ Ⓒ Ⓣ Ⓐ Ⓖ Ⓣ Ⓐ Ⓒ

23.32 Hydrogen bonding is responsible for holding together the two DNA strands. This is illustrated on Pages 765 and 766.

23.33 During cell division the two DNA strands unravel giving two complementary chains that serve as templates for the construction of two new daughter chains. The unraveled strand then begins to pair in such a way (T with A and C with G) that the two new strands are identical to the old one.

23.34 The genetic code is the set of base sequences that allow the tRNA to construct a protein with the correct amino acid sequence. Each amino acid in a polypeptide is specified by one or more sets of three bases. tRNA deciphers the code and brings amino acids to their proper positions along the mRNA. mRNA carries the genetic code from the DNA template within the nucleus to the ribosomes outside.

23.35 CGU—UUA—AAA—GGU—UGU. (There are other base sequences that could be chosen from Table 23.3.)

23.36

Ⓤ Ⓤ Ⓐ Ⓖ Ⓒ Ⓐ Ⓤ Ⓒ Ⓒ Ⓖ Ⓐ Ⓒ Ⓐ Ⓤ Ⓒ

23.37   Ser-Ile-Leu-Ser-Asn

23.38

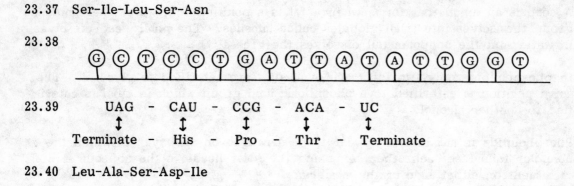

23.39   UAG - CAU - CCG - ACA - UC

        ↕   ↕   ↕   ↕   ↕

    Terminate - His - Pro - Thr - Terminate

23.40   Leu-Ala-Ser-Asp-Ile

23.41   Diseases arising from an altering of the base sequence in a DNA molecule. It is not always fatal to the cell.

CHAPTER 24

NUCLEAR CHEMISTRY

Rationale

There always seems to be considerable disagreement over the proper location of the discussion of nuclear transformations in a general chemistry text. The importance of nuclear sources of energy suggests that it should be given high priority. On the other hand, nuclear phenomena have very little direct influence on the outcome of chemical reactions, which suggests a low priority. Obviously, we have given in to the latter argument in determining the placement of this topic. In structuring the chapter, however, we have attempted to provide sufficient flexibility so that it can be discussed at almost any point after Chapter 3. Only the discussions of the kinetics of radioactive decay (p 775) and the application of tracer techniques to the investigation of reaction mechanisms (p 788) might cause difficulty because they require some background from Chapter 12. In fact, if you do anticipate moving the discussion of this chapter to an earlier point in the course, you may wish to treat these latter two topics within Chapter 12.

The chapter opens with an examination of the nuclear decay process, its kinetics, and applications to archaeological dating. This is followed by discussions of nuclear transformations, brought about by bombarding target nuclei with appropriate particles, and of nuclear stability. These lead easily into current and proposed extensions of the periodic table beyond the heaviest naturally occurring element, uranium.

Since this is a chemistry text, it seems appropriate that this chapter should contain a section dealing with chemical applications. Section 24.5 begins with a discussion of the chemical effects of radiation, followed by examples of applications of tracer techniques to chemical problems.

The final two sections consider nuclear reactions as sources of energy. Nuclear fission and fusion are examined in Section 24.6 from the point of view of the nuclear reaction involved. Section 24.7 traces the origin of the large energy changes accompanying these reactions.

Objectives

At the conclusion of this chapter, students should be able to:

Complete and balance nuclear equations given all but one reactant or one product.

Apply the kinetics of radioactive decay to the calculation of half-lives and to illustrative problems dealing with archaeological dating.

Define: nuclide, half-life, nuclear decay, nuclear transformations, band of stability, electron capture, fission, magic numbers, isotope dilution, neutron activation analysis, breeder reactor, nuclear fusion, critical mass, binding energy, mass defect.

State what characteristics are possessed by stable and unstable nuclei.

Give reactions that tend to bring unstable nuclei into the band of stability.

Explain how "magic numbers" predict the stability of superheavy elements.

Explain the principle of operation of: the Geiger Müller counter, the cyclotron, a breeder reactor.

Give examples of chemical applications of nuclear reactions.

Describe how nuclear fission can be harnessed to provide usable energy.

Perform calculations dealing with nuclear binding energy and mass defect.

Answers to Questions

24.1   Alpha particles, beta particles and gamma rays.

24.2   They are both particles with a mass equal to that of the electron, but they are opposite in charge.

24.3   (a) $^{81}_{36}Kr + ^{0}_{-1}e \longrightarrow ^{81}_{35}Br$ (c) $^{73}_{31}Ga \longrightarrow ^{0}_{-1}e + ^{73}_{32}Ge$

(b) $^{104}_{47}Ag \longrightarrow ^{0}_{1}e + ^{104}_{46}Pd$ (d) $^{104}_{48}Cd \longrightarrow ^{104}_{47}Ag + ^{0}_{1}e$

(e) $^{54}_{25}Mn + ^{0}_{-1}e \longrightarrow ^{54}_{24}Cr$

24.4 (a) $^{47}_{20}Ca \longrightarrow ^{47}_{21}Sc + ^{0}_{-1}e$

(d) $^{54}_{26}Fe + ^{1}_{0}n \longrightarrow ^{1}_{1}H + ^{54}_{25}Mn$

(b) $^{55}_{27}Co \longrightarrow ^{55}_{26}Fe + ^{0}_{1}e$

(e) $^{46}_{20}Ca + ^{1}_{0}n \longrightarrow ^{47}_{20}Ca$

(c) $^{220}_{86}Rn \longrightarrow ^{116}_{84}Po + ^{4}_{2}He$

24.5 (a) $^{135}_{53}I \longrightarrow ^{135}_{54}Xe + ^{0}_{-1}e$

(d) $^{96}_{42}Mo + ^{2}_{1}H \longrightarrow ^{1}_{0}n + ^{97}_{43}Tc$

(b) $^{245}_{97}Bk \longrightarrow ^{4}_{2}He + ^{241}_{95}Am$

(e) $^{20}_{8}O \longrightarrow ^{20}_{9}F + ^{0}_{-1}e$

(c) $^{238}_{92}U + ^{12}_{6}C \longrightarrow ^{246}_{98}Cf + 4\,^{1}_{0}n$

24.6 (a) $^{35}_{17}Cl + ^{1}_{0}n \longrightarrow ^{35}_{16}S + ^{1}_{1}H$

(d) $^{229}_{90}Th \longrightarrow ^{4}_{2}He + ^{225}_{88}Ra$

(b) $^{40}_{19}K \longrightarrow ^{0}_{-1}e + ^{40}_{20}Ca$

(e) $^{184}_{80}Hg \longrightarrow ^{184}_{79}Au + ^{0}_{1}e$

(c) $^{98}_{42}Mo + ^{1}_{0}n \longrightarrow ^{0}_{-1}e + ^{99}_{43}Tc$

24.7 (a) $^{11}_{5}B \longrightarrow ^{4}_{2}He + ^{7}_{3}Li$

(e) $^{116}_{51}Sb + ^{0}_{-1}e \longrightarrow ^{116}_{50}Sn$

(b) $^{90}_{38}Sr \longrightarrow ^{0}_{-1}e + ^{90}_{39}Y$

(f) $^{70}_{33}As \longrightarrow ^{0}_{1}e + ^{70}_{32}Ge$

(c) $^{107}_{47}Ag + ^{1}_{0}n \longrightarrow ^{108}_{47}Ag$

(g) $^{41}_{19}K \longrightarrow ^{1}_{1}H + ^{40}_{18}Ar$

(d) $^{88}_{35}Br \longrightarrow ^{1}_{0}n + ^{87}_{35}Br$

24.8 (a) $^{27}_{13}Al + ^{4}_{2}He \longrightarrow ^{1}_{0}n + ^{30}_{15}P$

(c) $^{15}_{7}N + ^{1}_{1}H \longrightarrow ^{4}_{2}He + ^{12}_{6}C$

(b) $^{209}_{83}Bi + ^{2}_{1}H \longrightarrow ^{1}_{0}n + ^{210}_{84}Po$

(d) $^{12}_{6}C + ^{1}_{1}H \longrightarrow ^{13}_{7}N + \gamma$

(e) $^{14}_{7}\text{N} + ^{4}_{2}\text{He} \longrightarrow ^{1}_{1}\text{H} + ^{17}_{8}\text{O}$

24.9 (a) $^{242}_{96}\text{Cm} + ^{4}_{2}\text{He} \longrightarrow ^{245}_{98}\text{Cf} + ^{1}_{0}\text{n}$ (d) $^{27}_{13}\text{Al} + ^{2}_{1}\text{H} \longrightarrow ^{25}_{12}\text{Mg} + ^{4}_{2}\text{He}$

(b) $^{108}_{48}\text{Cd} + ^{1}_{0}\text{n} \longrightarrow ^{109}_{48}\text{Cd} + \gamma$ (e) $^{249}_{98}\text{Cf} + ^{18}_{8}\text{O} \longrightarrow ^{263}_{106}\text{X} + 4\,^{1}_{0}\text{n}$

(c) $^{14}_{7}\text{N} + ^{1}_{0}\text{n} \longrightarrow ^{14}_{6}\text{C} + ^{1}_{1}\text{H}$

24.10 See Figure 24.1.

24.11

$$\log \frac{[A]_o}{[A]} = \frac{kt}{2.30} \qquad \text{(Equation 24.1)}$$

$$[A] = 1/2\,[A]_o \qquad @\ t_{1/2}$$

$$\log \frac{[A]_o}{1/2[A]_o} = \frac{kt_{1/2}}{2.30}$$

$$\log (2) = \frac{kt_{1/2}}{2.30} \qquad \text{or}$$

$$t_{1/2} = \frac{\log (2) \times 2.30}{k}$$

$$t_{1/2} = \frac{0.693}{k} \qquad \text{(Equation 24.2)}$$

24.12 As the number of protons in the nucleus increases there must be more and more neutrons present to help overcome the strong repulsive forces between the protons. Also, there seems to be an upper limit to the number of protons that can exist in a stable nucleus, that number being reached with bismuth. Nuclides above the band of stability must either lose neutrons or gain protons in order to achieve stability.

24.13 Elements higher than 83 must lose both neutrons and protons to achieve a stable n/p ratio. The only way this is possible is by α-emission or fission.

24.14 Nuclei that contain certain specific numbers of protons and neutrons possess a degree of extra stability. For protons these magic numbers are 2, 8, 20, 28, 50 and 82; for neutrons, 2, 8, 20, 28, 50, 82 and 126. The magic numbers for orbital electrons are 2, 8, 18, 36, and 54 (the number of electrons in closed electron shells).

24.15 e = even, o = odd, * = magic number

$$_2^4He > {}_{28}^{58}Ni > {}_{20}^{39}Ca > {}_{32}^{71}Ge > {}_5^{10}B$$

(e*,e*) (e*,e) (e*,o) (e,o) (o,o)

24.16 e = even, o = odd, * = magic number

$$_{77}^{192}Ir < {}_6^{13}C < {}_2^3He < {}_{50}^{116}Sn < {}_{20}^{40}Ca$$

(o,o) (e,o) (e*,o) (e*,e) (e*,e*)

24.17 Element 114 would fall under lead in Group IVA. It therefore would be a soft metal with a relatively low melting point. Its most stable oxidation state would be 2+ and would form such compounds as XO and XCl_2. A likely spot to discover this element would be wherever lead ores are found.

24.18 (a) Na_2X (b) H_2X (c) XO_2 (d) Since Po is a metalloid, 116 would probably have metallic properties.

24.19 298 and 310

24.20 Radiation emitted in quantum packets can be used to explain nuclear shells just as Bohr did in explaining his atomic theory. If protons move from shell to shell in the nucleus, each transition would result in an emission of energy.

24.21 The bombarding nuclei have to contain a very large n/p ratio to place the products on the island of stability (Figure 24.7). Light nuclei, however, contain n/p ratios of nearly 1.

24.22 Both Tc and Pm have an odd number of protons.

24.23 ON + ON*O   bond breaking at ①

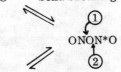

ONON*O

ONO + N*O   bond breaking at ②

24.24 Approximately half of the CH_3HgI should contain the labeled Hg. Two molecules of CH_3HgI could form $(CH_3)_2Hg$ plus HgI_2 when the reaction proceeds in the reverse direction. When the $(CH_3)_2Hg$ reacts in the forward direction, it can combine with either labeled or unlabeled HgI_2 producing labeled CH_3^*HgI.

24.25 One possible experiment would be to make the complex and allow the racemization to occur in a medium containing labeled $C_2O_4^{2-}$. If the racemization occurs by the dissociation of a $C_2O_4^{2-}$, the complex should pick up some labeled $C_2O_4^{2-}$.

LECTURE DEMONSTRATIONS

The following pages provide a list of some demonstrations that you might consider performing in your classes. References are to Tested Demonstrations in Chemistry, edited by Hubert N. Alyea and Frederick B. Dutton, and published by the Division of Chemical Education of the American Chemical Society, Easton, Pennsylvania. Demonstrations that are not referenced are others that have been performed by the Author. Please note that space limitations prevent detailed descriptions of procedures. Those performing any of these demonstrations should be sure that they are fully aware of any potential hazards that might exist when the demonstrations are carried out.

CHAPTER 1

INTRODUCTION

Measurement

Display meter stick, 1-liter flask or beaker filled with water, and one kilogram weight. Perhaps also some common products that report quantities using metric units.

Physical and Chemical Changes

"Tested Demonstrations," Page 5, Parts C and D.

Mixtures

"Tested Demonstrations," Page 5, Part B.

Energy

"Tested Demonstrations," Page 17, Parts 7-1, 7-4, 7-5, 7-9, 7-10, 7-13.

CHAPTER 2

STOICHIOMETRY: CHEMICAL ARITHMETIC

The Mole

Exhibit 1 mol of various substances such as copper pennies, iron nails, sodium chloride, sodium bicarbonate, water. Point out that each sample contains the same number of formula units.

Solution Preparation

Demonstrate how a volumetric flask is used to prepare a solution of a particular molarity.

CHAPTER 3

ATOMIC STRUCTURE AND THE PERIODIC TABLE

Gas Discharge Tubes

A high voltage vacuum leak detector can be used to cause a fluorescent lamp to light. You can explain how the electric discharge causes mercury vapor in the tube to emit ultraviolet light which is converted to visible light by the phosphor that coats the inner wall of the tube.

Atomic Spectra

Perform flame tests on solutions of LiCl, NaCl, KCl, $CaCl_2$, $SrCl_2$, $BaCl_2$, and $CuCl_2$. Discuss line spectra and their significance.

Waves and Harmonics

Use a guitar to demonstrate that only certain wavelengths are permitted on an open string. Play open string and then various harmonics. This leads naturally into the concept of standing waves and an easy extension to the discussion of electron waves in atoms.

Paramagnetism

"Tested Demonstrations," Page 215.

CHAPTER 4

CHEMICAL BONDING: GENERAL CONCEPTS

Formation of an Ionic Compound

"Tested Demonstrations," Page 43, Part 20-6: Reaction of Mg with Cl_2.

Formation of a Covalent Compound

"Tested Demonstrations," Page 43, Part 20-8.

CHAPTER 5

COVALENT BONDING AND MOLECULAR STRUCTURE

Shapes of Molecules

Use molecular models to demonstrate the basic molecular geometries. These can also be used in the discussion of the VSEPR theory.

Hybrid Orbitals

The Atomic Molecular Orbital Models by Science Related Materials, Inc., Janesville, Wisconsin, are useful for illustrating the orientations of the hybrid orbitals and the formation of double and triple bonds by orbital overlap.

CHAPTER 6

CHEMICAL REACTIONS IN AQUEOUS SOLUTION

Introduction

Demonstrate a reaction in aqueous solution by slowly adding a solution of NaCl to a solution of $AgNO_3$. Write the chemical equation for the reaction.

Electrolytes

Use a conductivity apparatus like that illustrated in Figure 6.3 to test the conductivity of (a) distilled water, (b) a NaCl solution, (c) a sugar solution, (d) glacial acetic acid, then gradually add water to the acetic acid while the electrodes are immersed in the acid, (e) concentrated aqueous ammonia, (f) dilute hydrochloric acid.

Ionic Reactions

Demonstrate again the reaction between NaCl and $AgNO_3$. Using the conductivity apparatus, add glacial acetic acid to concentrated aqueous ammonia. Using the conductivity apparatus, observe what happens as dilute sulfuric acid is added gradually to a solution of $Ba(OH)_2$. Add a solution of HCl to a solution of Na_2CO_3. Write molecular, ionic, and net ionic equations for the reactions that occurred.

Precipitation Reactions

"Tested Demonstrations," Page 202.

Redox Reactions

Add a solution of $KMnO_4$ slowly to a solution of oxalic acid acidified with H_2SO_4. Set up and balance the net ionic equation for the reaction using the ion-electron method. Explain why $KMnO_4$ is a useful reagent for titrations in redox reactions.

CHAPTER 7
GASES

Atmospheric Pressure

Use a vacuum pump to evacuate an empty, dry 5-gallon solvent can. Atmospheric pressure causes the can to collapse.

Gas Laws

"Tested Demonstrations," Page 64.

Graham's Law

"Tested Demonstrations," Page 188, 204.

Kinetic Molecular Theory

"Tested Demonstrations," Page 153.

CHAPTER 8
STATES OF MATTER AND INTERMOLECULAR FORCES

Surface Tension

"Tested Demonstrations," Page 189.

Liquid-Vapor Equilibrium

"Tested Demonstrations," Page 180.

Sublimation

To show the sublimation of naphthalene, place a layer of the solid about 5 mm thick at the bottom of a 400-ml beaker and place the beaker on a wire gauze on an iron ring mounted on a ring stand. Cover the beaker with a large watch glass and fill the depression in the watch glass with ice. Warm the beaker. Beautiful, delicate crystals of naphthalene will form on the underside of the watch glass. If possible, illuminate the crystals with a strong light, for example, using a 35-mm projector, so they glisten as you lift the watch glass.

Heat of Crystallization

"Tested Demonstrations," Page 187.

CHAPTER 9

THE PERIODIC TABLE REVISITED

Physical Properties of Metals and Nonmetals

Exhibit lumps of sulfur and a length of copper wire. Comment on the physical appearance of the two elements. Show how the sulfur crystals crumble when struck by a hammer and how the copper is malleable.

Ease of Oxidation of Metals

Add pieces of magnesium, zinc, iron (nails), and copper to dilute hydrochloric acid in separate 8-inch test tubes. Note the relative rates of reaction. Place copper in nitric acid to show how it dissolves in that acid but not in HCl.

Colors of Metal Compounds

Add a solution of silver nitrate to solutions of NaCl, NaBr, and NaI in separate test tubes. Note the colors of the precipitates and discuss the relative degrees of covalent character in the Ag-X bonds.

CHAPTER 10
PROPERTIES OF SOLUTIONS

Saturation and Supersaturation

"Tested Demonstrations," Part 5-21, Page 14.

Polar Properties and Solubility

"Tested Demonstrations," Page 223.

Effect of Temperature on Solubility

"Tested Demonstrations," Page 189.

Raoult's Law

"Tested Demonstrations," Page 195.

CHAPTER 11
CHEMICAL THERMODYNAMICS

Spontaneous and Nonspontaneous Reactions

Collect a test tube full of hydrogen and demonstrate the spontaneous reaction of H_2 and O_2 to form H_2O by igniting the mixture. Then set up an apparatus for the electrolysis of water (3 M H_2SO_4 as the electrolyte). Use several dry cells connected in series as the source of electricity. Show that the electrolysis only proceeds as long as the electricity is provided by the spontaneous reactions taking place inside the batteries.

Entropy and Spontaneity

Place a drop of concentrated $KMnO_4$ solution carefully into a 600-ml beaker of still water. Note over a period of time how the solute gradually diffuses spontaneously throughout the water.

CHAPTER 12
CHEMICAL KINETICS

Effect of Concentration and Temperature on Reaction Rate

"Tested Demonstrations," Part A, Page 19. Other clock reactions are described on Pages 85 and 179.

Catalysis

"Tested Demonstrations," Page 159.

CHAPTER 13

CHEMICAL EQUILIBRIUM

Le Châtelier's Principle

"Tested Demonstrations," Part C, Page 19; Part C, Page 86; Page 167; Page 221.

CHAPTER 14

ACIDS AND BASES

Brønsted Acids and Bases

Remove the caps from bottles of concentrated HCl and NH_3, and hold the mouths of the bottles near each other. A cloud of ammonium chloride forms. Point out that this is an acid-base reaction occurring in the absence of a solvent.

Lewis Acids and Bases

"Tested Demonstrations," Page 194.

CHAPTER 15

ACID-BASE EQUILIBRIA IN AQUEOUS SOLUTION

Buffers

"Tested Demonstrations," Part 4-32s, Page 128. Also Page 155.

Hydrolysis of Salts

"Tested Demonstrations," Part 4-28s, Page 62.

182

CHAPTER 16
SOLUBILITY AND COMPLEX ION EQUILIBRIA

Solubility Product

"Tested Demonstrations," Part 8-15s, Page 86.

Complex Ions and Solubility

To a solution of silver nitrate in a 600-ml beaker, slowly add a concentrated solution of sodium iodide. At first a precipitate of AgI is formed. As additional I^- is added, the precipitate dissolves as the complex ion AgI_2^- is formed. Dilution of the solution containing the complex ion causes the AgI to reprecipitate. Explain the observations in terms of the equilibria involved.

CHAPTER 17
ELECTROCHEMISTRY

Electrolysis

"Tested Demonstrations," Page 161.

Galvanic Cells

"Tested Demonstrations," Page 150 and Page 154.

CHAPTER 18
CHEMICAL PROPERTIES OF THE REPRESENTATIVE METALS

The Alkali Metals

"Tested Demonstrations," Page 27.

The Alkaline Earth Metals

"Tested Demonstrations," Page 29.

Reaction of Magnesium with Steam

"Tested Demonstrations," Page 142.

Thermite Reaction

"Tested Demonstrations," Page 17.

Amphoteric Nature of Aluminum

Roll two 5-cm x 5-cm squares of aluminum foil into balls small enough to fit into the mouth of an 8-inch test tube. Half fill an 8-inch test tube with 6 M HCl; half fill another with 6 M NaOH. Drop an aluminum foil ball into each solution. When they are reacting vigorously, use a match to ignite the hydrogen gas being given off in each. (Note: sometimes it helps to cautiously warm the contents of the test tubes to get the reactions started.)

CHAPTER 19

THE CHEMISTRY OF SELECTED NONMETALS, PART I:

HYDROGEN, CARBON, OXYGEN, AND NITROGEN

Hydrogen

"Tested Demonstrations," Page 9.

Carbon

"Tested Demonstrations," Page 35.

Oxygen

"Tested Demonstrations," Page 7.

Nitrogen

"Tested Demonstrations," Page 37. In demonstrating the Ostwald Process, the apparatus shown in Color Plate 16 can be used. The catalyst pictured here is a platinum electrode. To work effectively, the platinum surface must be cleaned thoroughly.

CHAPTER 20

THE CHEMISTRY OF SELECTED NONMETALS, PART II:

PHOSPHORUS, SULFUR, THE HALOGENS, THE NOBLE GASES, AND SILICON

Sulfur

"Tested Demonstrations," Page 41.

The Halogens

"Tested Demonstrations," Page 43 and Page 213.

184

Silicon

"Tested Demonstrations," Page 33.

CHAPTER 21
THE TRANSITION ELEMENTS

Metallurgy

"Tested Demonstrations," Page 25.

Chromate-Dichromate Equilibrium

"Tested Demonstrations," Page 182.

Complex Ions

The colors and formation of complex ions can be illustrated by a few very simple experiments. Aqueous ammonia added to Cu^{2+} gives the blue $Cu(NH)_3{}^{2+}$ ion. Addition of NCS^- to a solution containing Fe^{3+} gives a red iron(III) thiocyanate complex.

CHAPTER 22
ORGANIC CHEMISTRY

Properties of Organic Compounds

"Tested Demonstrations," Page 47.

Preparation of Nylon

"Tested Demonstrations," Page 164.

CHAPTER 23
BIOCHEMISTRY

Carbohydrates

Dehydration of sugar using concentrated sulfuric acid demonstrates that the overall composition of carbohydrates is $C_n(H_2O)_m$.

Lipids and Soap

"Tested Demonstrations," Part 22-15, Page 48.

CHAPTER 24
NUCLEAR CHEMISTRY

Radioactivity

"Tested Demonstrations," Page 21.

STEREOSCOPIC ILLUSTRATIONS

The following pages contain most of the stereoscopic illustrations that appeared in the second edition of the textbook. John Wiley and Sons gives you permission to reproduce them in whatever quantity necessary for use by your students. In order to obtain a three-dimensional illusion from them, most people require the use of a viewer. Viewers can be purchased in quantities of ten or more by writing to:

> Taylor-Merchant Corporation
> 25 West 45th Street
> New York, N. Y. 10036

For those who have not had occasion to view these kinds of illustrations, you will notice that each consists of a pair of drawings or photographs that at first glance appear to be identical. Actually, they are slightly different. When viewed in such a way that the left eye focuses on the left drawing and the right eye focuses on the right drawing, your mind brings them together and creates a three-dimensional image. A viewer, like that sold by Taylor-Merchant assists you in obtaining a 3-D illusion.

To use the viewer, assemble it according to the directions printed on it and place the bottom edge of the viewer along the bottom of the box drawn around the illustration. The viewer should be placed so that the folded support panel is placed between the two drawings. Look through the lenses of the viewer keeping both eyes open. Start with your eyes a few inches above the viewer. At first you may find that it takes a moment for the stereo image to fuse. You may have to move the viewer slightly if a double image persists. With practice you will find that the stereo images become easier to see.

187

Figure 3.3
A cathode-ray tube used to measure the charge-to-mass ratio of the electron.

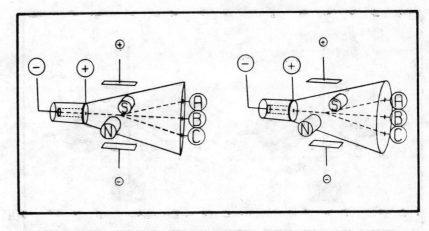

Figure 3.25
A three-dimensional representation of a 1s orbital. From D. T. Cromer, Journal of Chemical Education, Vol. 45, p. 626, October 1968, used by permission.

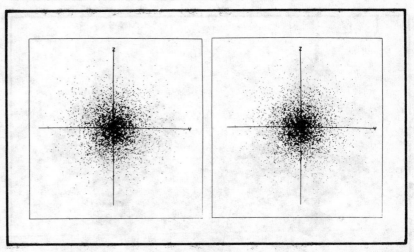

Figure 3.28
A three-dimensional representation of a 2p orbital. From D. T. Cromer, Journal of Chemical Education, Vol. 45, p. 626, October 1968, used by permission.

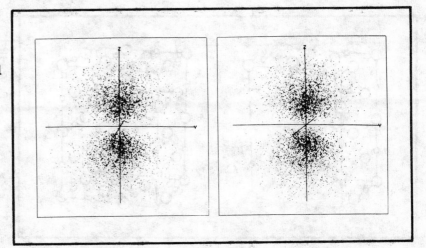

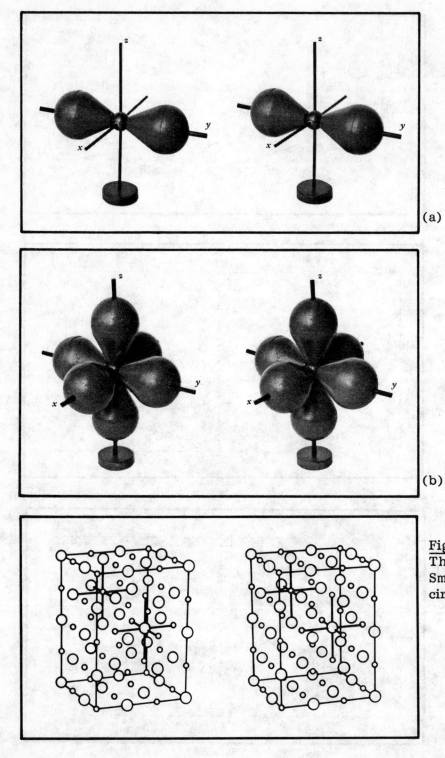

(a)

(b)

Figure 3.29
Stereo photographs showing the directional properties of p orbitals.
(a) A single p orbital (a p_y orbital).
(b) Three p orbitals (p_x, p_y, p_z) on a single set of axes.
Atomic-Molecular Orbital Models by Science Related Materials, Inc., Janesville, Wisconsin.

Figure 4.1
The structure of LiF.
Small circles = Li^+; large circles = F^-.

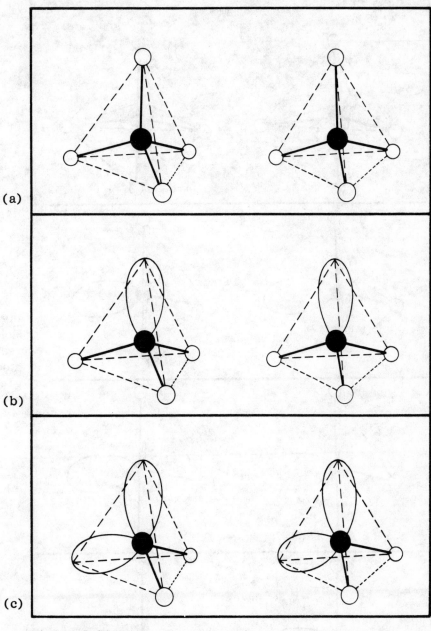

Figure 5.3
Geometries of molecules in which the central atom has four pairs of electrons. (a) AX_4 (for example, CH_4).
(b) AX_3E (for example, NH_3). (c) AX_2E_2 (for example, H_2O).

190

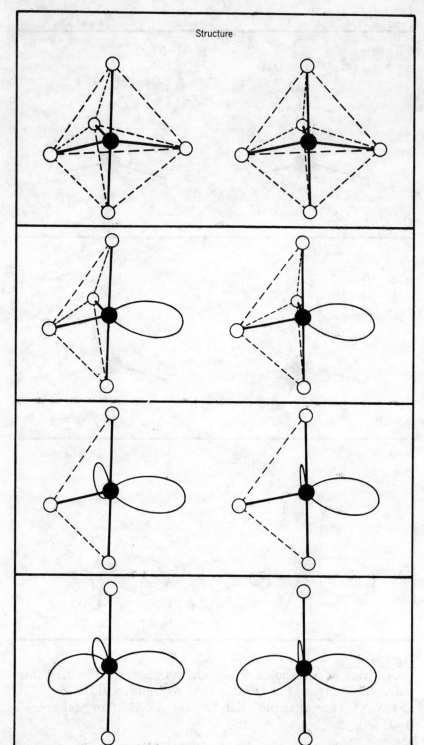

Structure

Figure 5.4
Molecular structures that
result when the central
atom has five electron-pair
groups.

Structure

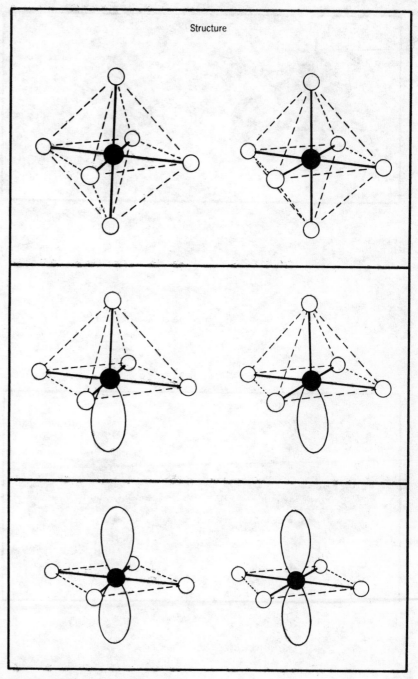

Figure 5.5
Molecular structures that result when the central atom has
six electron-pair groups.

192

Figure 5.7
Formation of HF by overlap
of partially filled fluorine
2p orbital with 1s orbital of
hydrogen. Heavy shading
indicates filled orbital, light
shading indicates partially
filled orbital. Atomic-
Molecular Orbital Models by
Science Related Materials,
Inc., Janesville, Wisconsin.

Figure 5.8
Bonding in H_2O. Overlap
of two half-filled oxygen 2p
orbitals with the hydrogen
1s orbitals. Atomic-
Molecular Orbital Models by
Science Related Materials,
Inc., Janesville, Wisconsin.

Figure 5.9
Bonding in NH_3 gives a
pyramidal molecule.
(a) Overlap of 2p orbitals
of nitrogen with 1s orbitals
of hydrogen. (b) Pyramidal
shape of the NH_3 molecule.
Atomic-Molecular Orbital
Models by Science Related
Materials, Inc., Janesville,
Wisconsin.

(a)

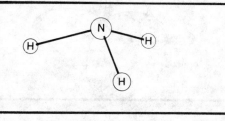

(b)

Figure 5.10
The structure of methane.
Solid sphere—carbon.

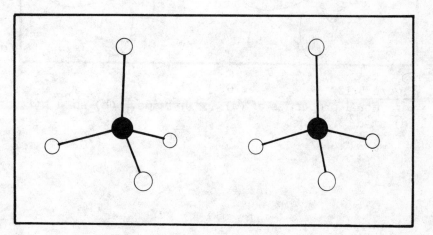

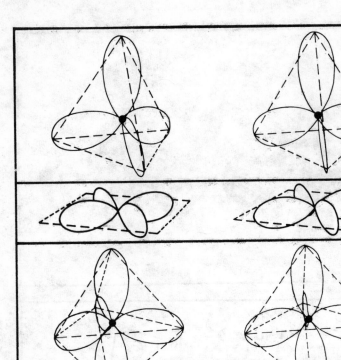

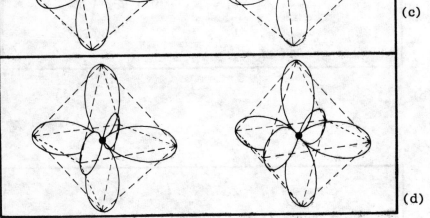

Figure 5.12
Directional properties of (a) sp^3 hybrids.   (b) sp^2d hybrids.   (c) sp^3d hybrids.
(d) sp^3d^2 hybrids.

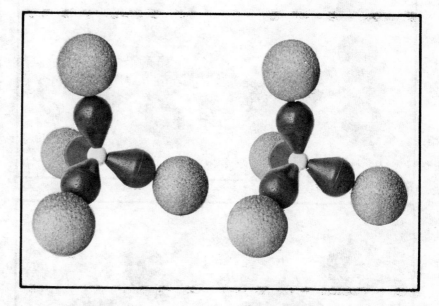

Figure 5.14
The formation of methane by overlap of hydrogen 1s orbitals with carbon sp^3 hybrids. Atomic-Molecular Orbital Models by Science Related Materials, Inc., Janesville, Wisconsin.

196

Figure 5.15
The use of sp³ hybrids for bonding in (a) H_2O and (b) NH_3. Spheres represent hydrogen 1s orbitals. Atomic-Molecular Orbital Models by Science Related Materials, Inc., Janesville, Wisconsin.

(a)

(b)

Figure 5.17
Sigma and pi bonds in ethylene. Atomic-Molecular Orbital Models by Science Related Materials, Inc., Janesville, Wisconsin.

197

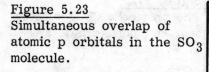

Figure 5.23
Simultaneous overlap of
atomic p orbitals in the SO_3
molecule.

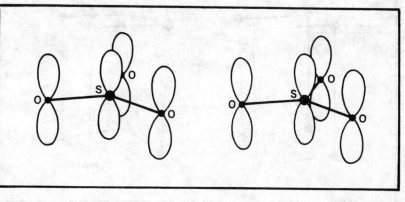

Figure 8.10
The crystal structure of ice
Large circles are oxygen.
Small circles represent hydrogen. Only half of the
small circles are occupied
by H atoms. From J. M.
Williams, Journal of Chemical Education, Vol. 52,
p. 210, April 1975, used
by permission.

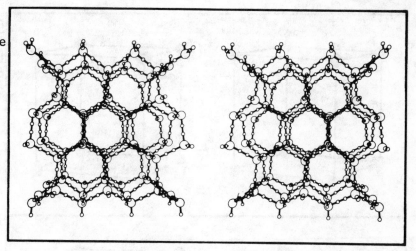

Figure 8.24
A simple cubic lattice.

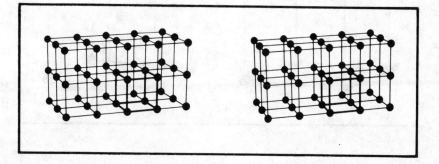

198

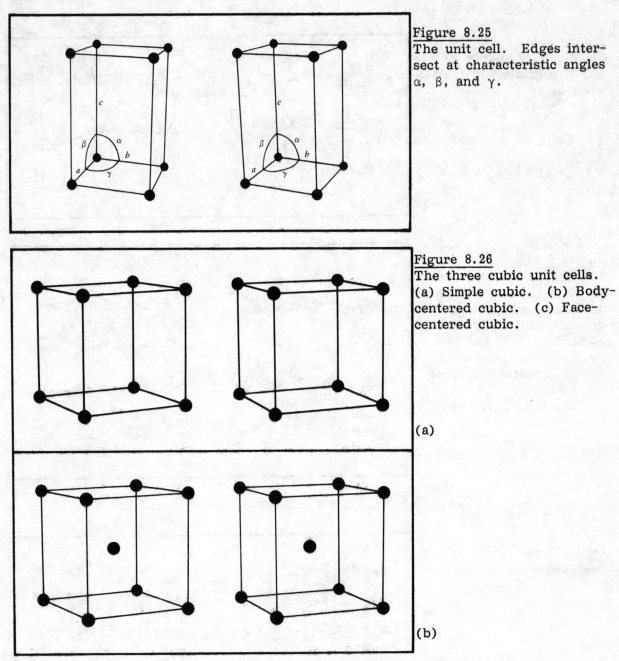

Figure 8.25
The unit cell. Edges intersect at characteristic angles α, β, and γ.

Figure 8.26
The three cubic unit cells.
(a) Simple cubic. (b) Body-centered cubic. (c) Face-centered cubic.

(a)

(b)

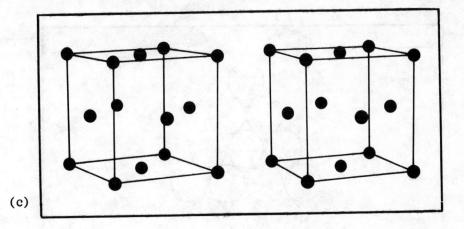

(c)

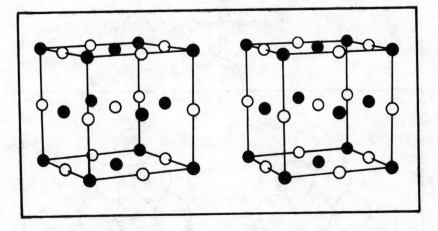

Figure 8.27
The sodium chloride
structure.

The zinc-blende structure.

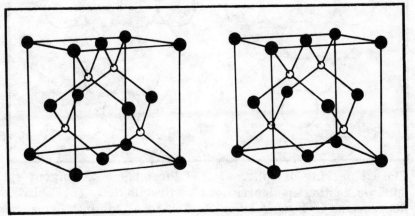

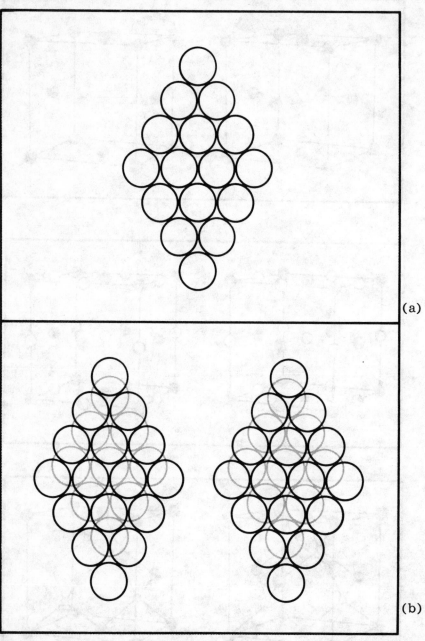

(a)

(b)

Closest packing of spheres. (a) First layer of tangent spheres. (b) Second layer of spheres resting in depressions in first layer. (c) Third layer of spheres resting in depressions in second layer over spheres in first layer—hcp structure. (d) Third layer of spheres resting in depressions in second layer over unused depressions in first layer—ccp structure.

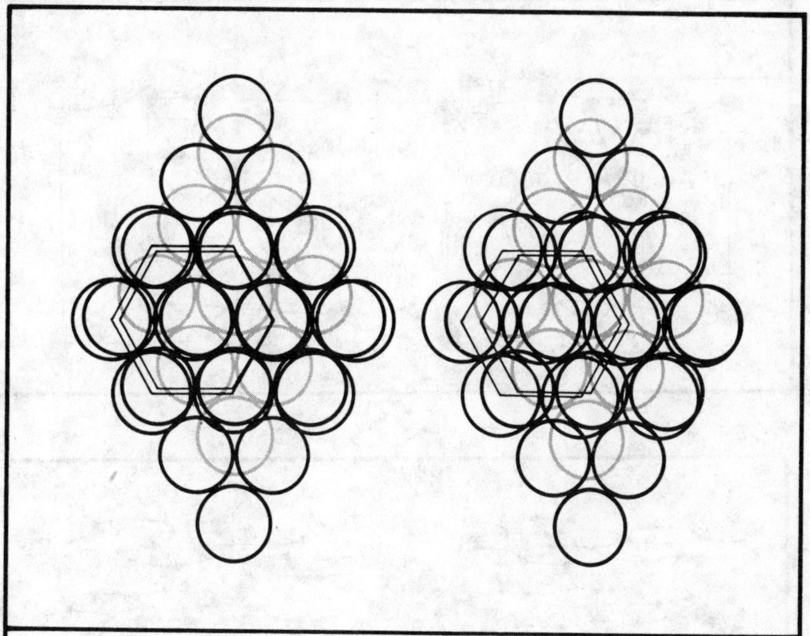

(c)

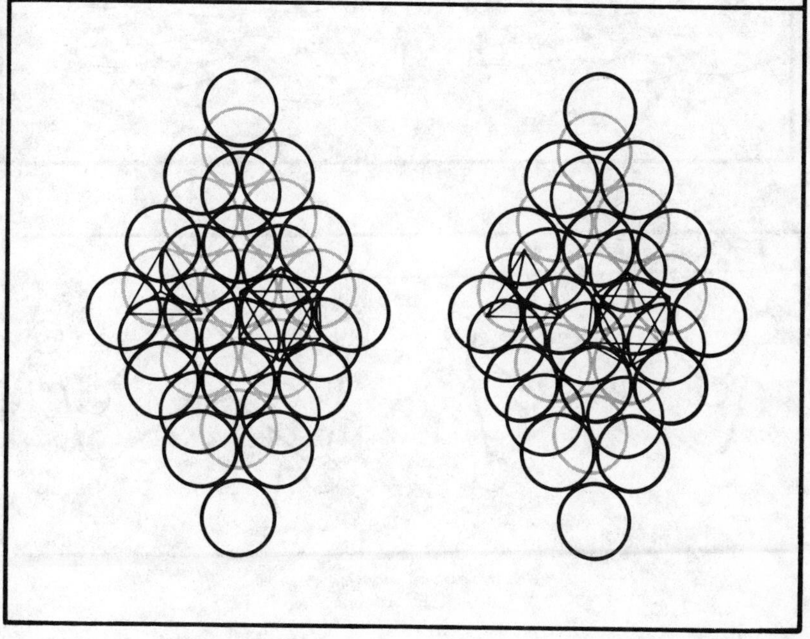

(d)

202

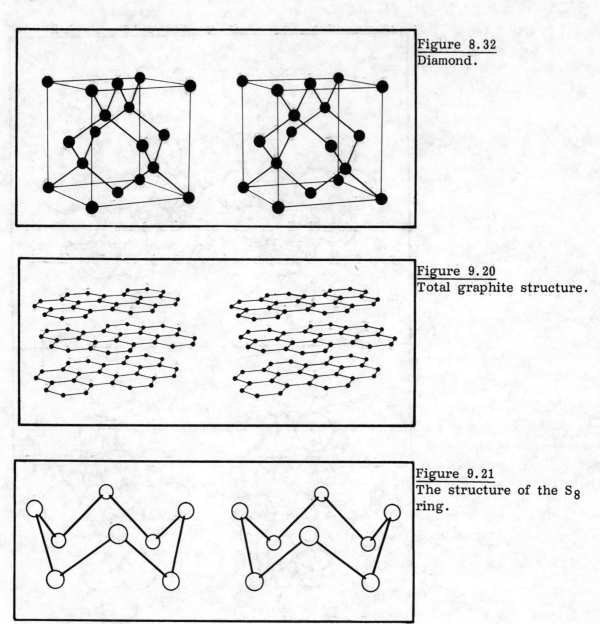

Figure 8.32
Diamond.

Figure 9.20
Total graphite structure.

Figure 9.21
The structure of the S_8 ring.

Figure 9.22
The structure of white
phosphorus, P_4.

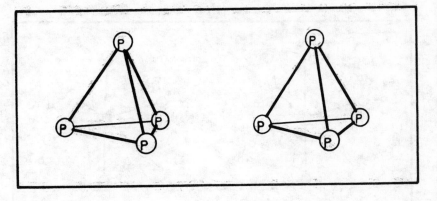

A layer in the structure
of black phosphorus.

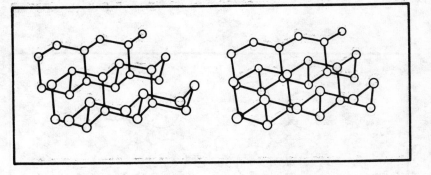

Figure 9.23
The icosahedral B_{12} unit
in elemental boron.

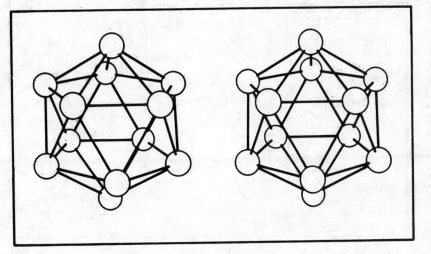

204

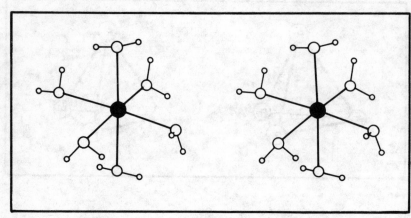

Figure 14.1
The hydrated aluminum ion, $Al(H_2O)_6^{3+}$. The aluminum ion (large solid sphere) is surrounded octahedrally by six water molecules.

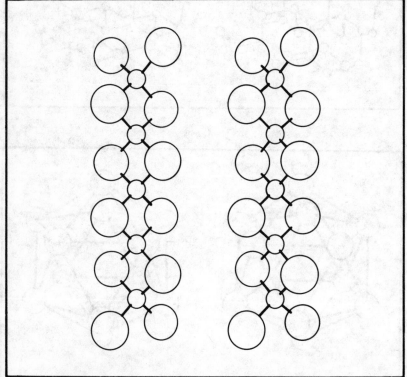

Figure 18.7
Structure of $(BeCl_2)_x$ (small spheres, Be; large spheres, Cl).

Structure of Al_2X_6 (small spheres, Al; large spheres, X).

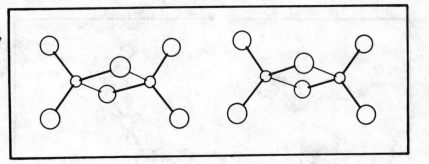

The structure of P_4O_{10}.

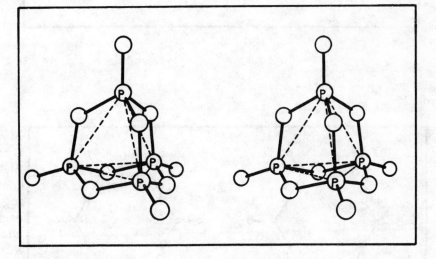

The structure of P_4O_6.

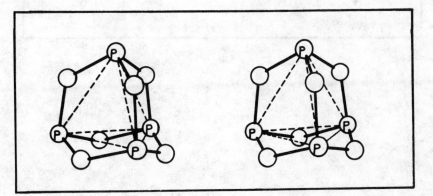

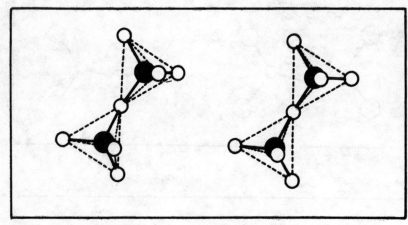

Figure 20.11
The structure of the $Si_2O_7^{6-}$ ion.

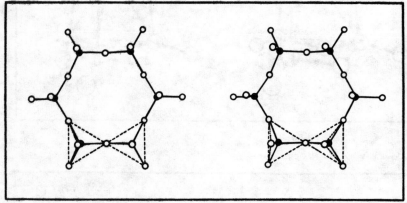

Figure 20.12
Cyclic structure of the $Si_6O_{18}^{12-}$ anion. Notice that there are 12 nonbridging oxygen atoms that each carry a single negative charge. (Solid spheres = X.)

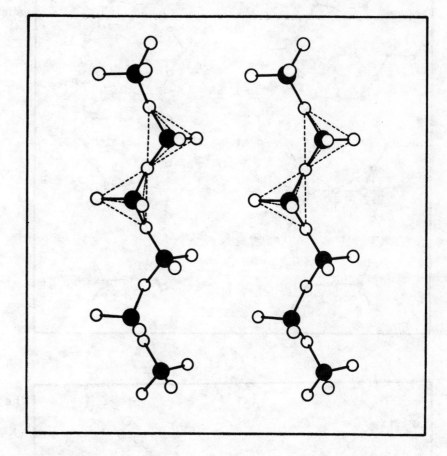

Figure 20.13
A segment of the linear SiO_3^- ion.

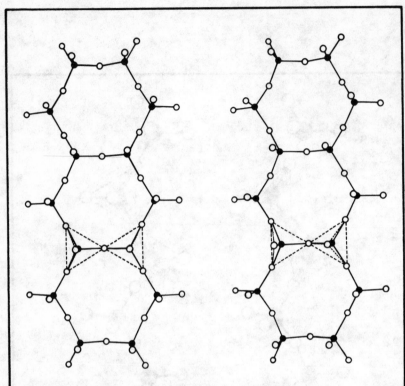

Figure 20.14
Linear double chain of SiO_4 tetrahedra that is found in asbestos.

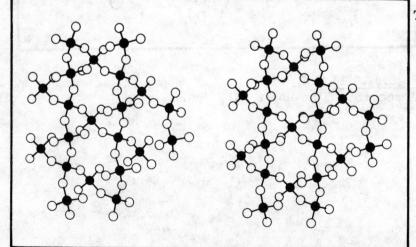

The structure of quartz.

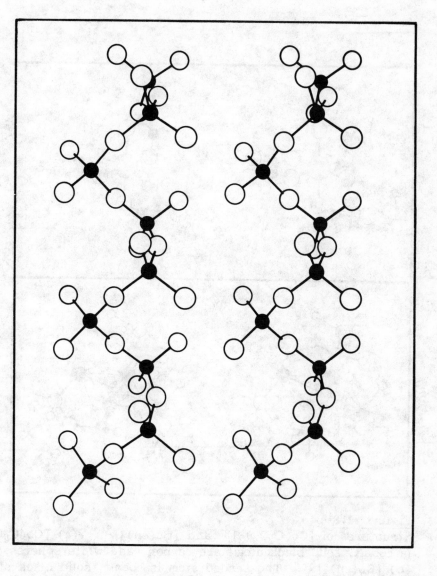

Spiral chains of SiO$_4$ tetrahedra found in quartz.

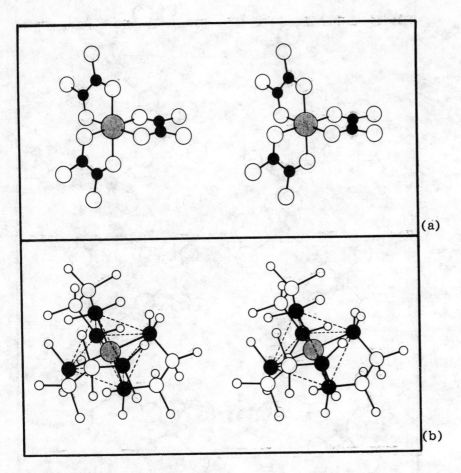

Figure 21.8
Structures of $[Co(C_2O_4)_3]^{3-}$ and $[Co(en)_3]^{3+}$. (a) $[Co(C_2O_4)_3]^{3-}$. The shaded atom is cobalt, solid black atoms are carbon, and white spheres are oxygen atoms. (b) $[Co(en)_3]^{3+}$. The shaded atom is cobalt, solid black atoms are nitrogen, the large white atoms are carbon, and the small white atoms are hydrogen.

211

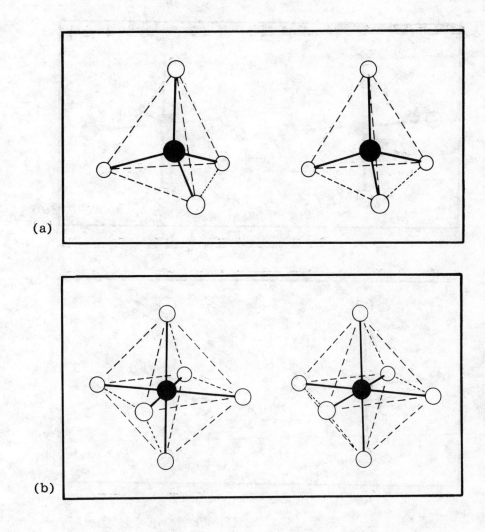

(a) Tetrahedral and (b) octahedral coordination.

212

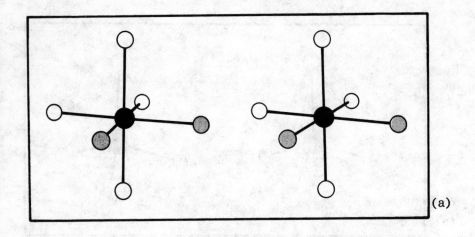

(a)

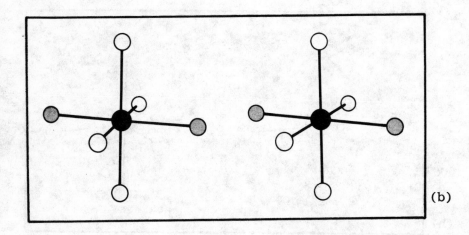

(b)

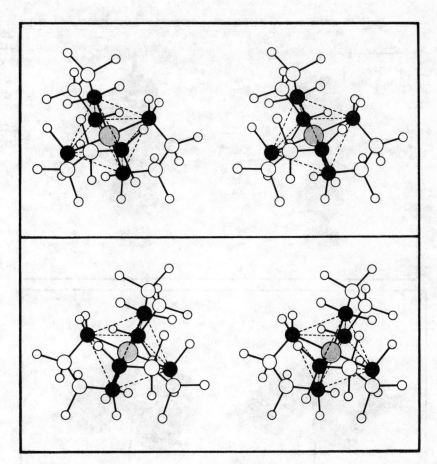

Figure 21.14
The complex $[Co(en)_3]^{3+}$.
The two structures shown
are nonsuperimposable
mirror images of one
another.

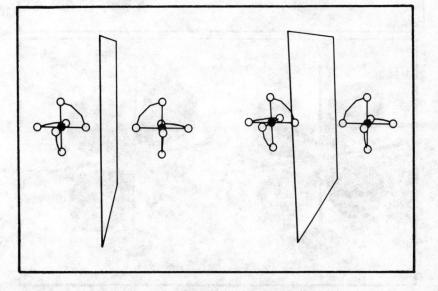

Mirror image relationship
between optical isomers of
$[Co(en)_3]^{3+}$.

214

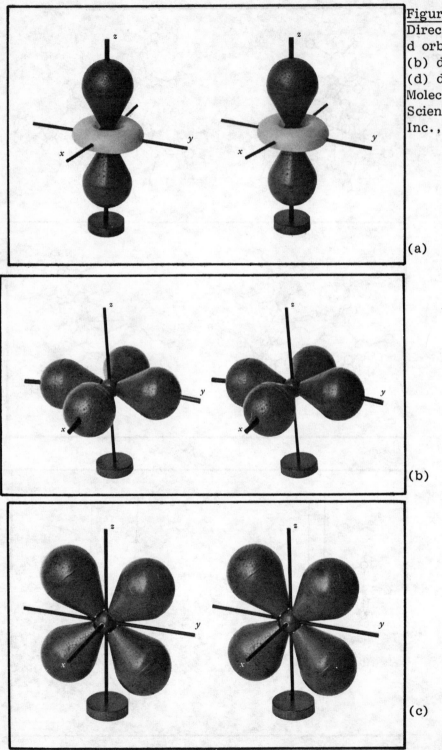

Figure 21.20
Directional properties of the d orbitals. (a) d_{z^2}. (b) $d_{x^2-y^2}$. (c) d_{yz}. (d) d_{xy}. (e) d_{xz}. Atomic-Molecular Orbital Models by Science Related Materials, Inc., Janesville, Wisconsin.

(a)

(b)

(c)

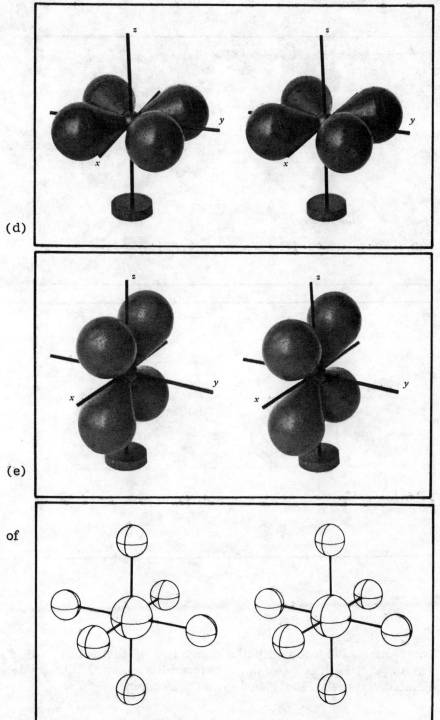

(d)

(e)

Figure 21.21
Octahedral arrangement of
ligands about a central
metal ion.

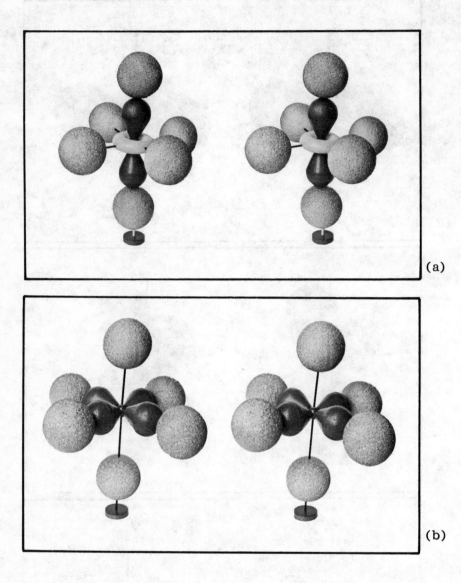

(a)

(b)

Figure 21.22
Interaction of ligands with the d orbitals of the metal. (a) d_{z^2}. (b) $d_{x^2-y^2}$
(c) d_{yz}. (d) d_{xz}. (e) d_{xy}. Atomic-Molecular Orbital Models by Science
Related Materials, Inc., Janesville, Wisconsin.

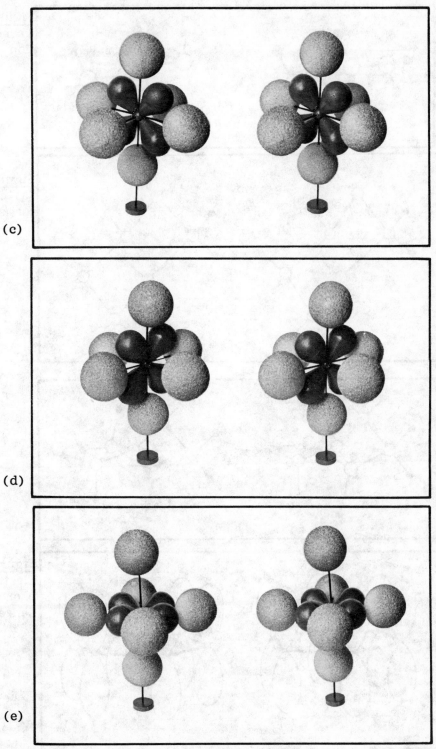

(c)

(d)

(e)

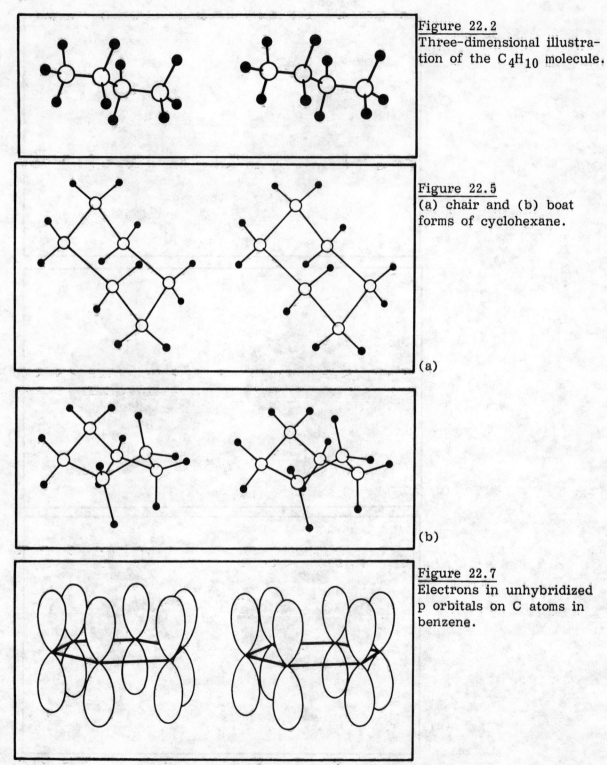

Figure 22.2
Three-dimensional illustration of the C_4H_{10} molecule.

Figure 22.5
(a) chair and (b) boat forms of cyclohexane.

(a)

(b)

Figure 22.7
Electrons in unhybridized p orbitals on C atoms in benzene.

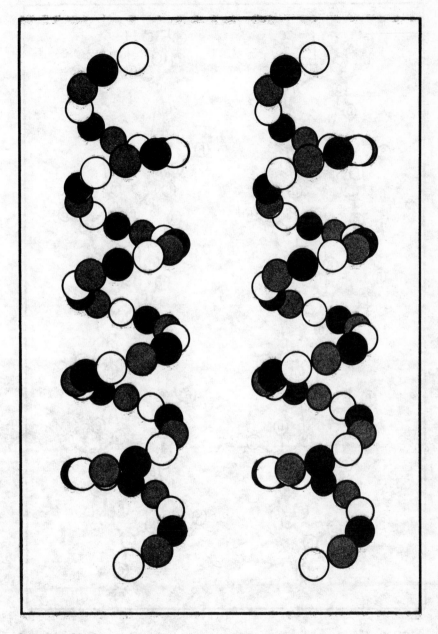

Figure 23.2
The α-helix. (See Page 747 of the text.)

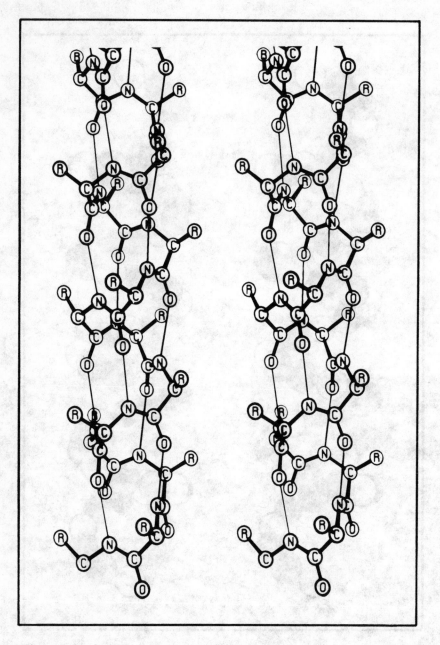

Figure 23.3
Hydrogen bonding (light lines) in the α-helix. Courtesy of Carroll K. Johnson, Oak Ridge National Laboratory, Oak Ridge, Tennessee.

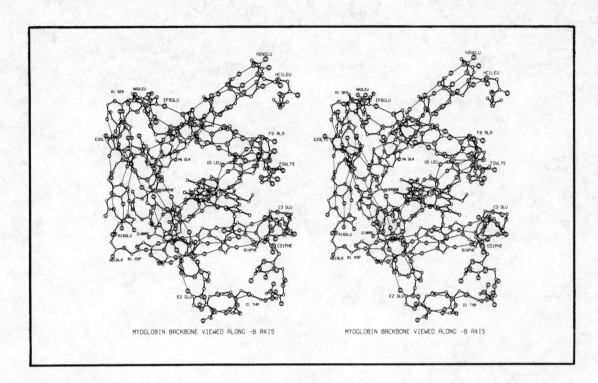

MYOGLOBIN BACKBONE VIEWED ALONG -B AXIS   MYOGLOBIN BACKBONE VIEWED ALONG -B AXIS

Figure 23.4
The tertiary structure of myoblobin. Courtesy of Carroll K. Johnson, Oak
Ridge National Laboratory, Oak Ridge, Tennessee.

Figure 23.9
Vitamin B-12 coenzyme.
Courtesy of Carroll K.
Johnson, Oak Ridge
National Laboratory,
Oak Ridge, Tennessee.

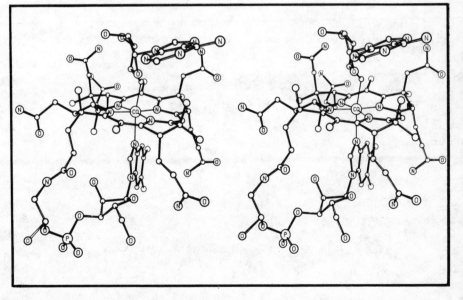